T0140306

Internet of Things and Smart Environments

Seyed Shahrestani

Internet of Things and Smart Environments

Assistive Technologies for Disability, Dementia, and Aging

 Springer

Seyed Shahrestani
Western Sydney University
Sydney, Australia

ISBN 978-3-319-86793-9   ISBN 978-3-319-60164-9   (eBook)
DOI 10.1007/978-3-319-60164-9

Printed on acid-free paper

This Springer imprint is published by Springer Nature
The registered company is Springer International Publishing AG
The registered company address is: Gewerbestrasse 11, 6330 Cham, Switzerland

To Fariba, Sara, and Sam.
Hope this is worth the time it has taken from
us.

Preface

Older adults and individuals with sensory or cognitive impairments face many barriers to inclusion and accessing various opportunities and services that the society has to offer. To overcome the disabling effects of these barriers, some of these people may employ some form of assistive technologies (AT). With their prevalence and diverse capabilities, the Internet of Things (IoT), smart homes, smart buildings, smart cities, and other smart environments can overcome some of the challenges that the more traditional AT face. There are many stimulating works on the use of assistive IoT and smart environments. This book, conscious of the patience of its potentially diverse range of readers, provides a current review and a sample of available services, techniques, concepts, and research work relevant to assistive IoT and smart environments.

By implanting sensors and actuators into their products and services, the enterprises have started to take advantage of the opportunities that the IoT offers. The IoT adoptions have also enabled the development of smart environments. These advances are drastically shifting the ways that people engage with the physical world. The pervasive nature of the IoT and the context-aware characteristics that stem from connecting many things together support the provisions of ambient intelligence. The capabilities to communicate such intelligence with other machines, including those nearby and those providing cloud-based deep learning and AI capacities, hold significant potentials for improving people's lives in many respects.

The IoT-based assistive systems have the possibility of extremely enhancing human experiences and quality of life. Smart environments can shift the focus away from the assistive devices or technology towards objects or things, putting the emphasis on how the environment can accommodate an individual's daily living requirements in the setups that they can construct for themselves. With their ubiquitous nature and explosive expansions, these technologies and infrastructure can play leading roles in revolutionizing aged care and empowering people with disability or dementia. However, this does not appear to be the case in practice. The issues relevant to the proliferation of the assistive IoT and smart environments will be discussed throughout this book. Improvements in many areas, particularly regarding the development of appropriate applications, implementations, and usability or adoptions, are required. A general aspect that needs particular attention relates to the requirement of taking a holistic approach that can provide or assist with improving the independence, access provisions, and integration of the concerned individuals while providing support for their families and caregivers. These are the areas that innovations and research are truly required. With the aim of motivating more investigations and developments, some of the challenges that prevent the widespread adoption of the assistive IoT-based solutions are also outlined in the book.

With over one billion people estimated to be living with disability, they have been considered as the world's largest minority. Aging has clear influences on the disability prevalence, as the potential of disability increases with getting older. Cognitive abilities vary widely among people, but they may also correlate with age. While some deteriorations of these abilities naturally occur with aging, some disorders may have more effects on them. Dementia is an example of such disorders. Dementia ensues from an acquired brain ailment, resulting in progressive deterioration of cognitive functions. To make any meaningful progress, the effects of aging, dementia, and disability on the individuals, their families and caregivers, social infrastructure, and economy need to be understood. As such, an overview of the scale, growth rates, and demography of disability, dementia, and aging is part of this book. This overview is complemented with discussions on the individual and societal effects of age and disability, along with their impacts on family members and caregivers, as well as on community, economy, and national expenditure. The dominant age and disability views and models along with their effects on AT developments and adoptions are also presented.

This book is focused on identifying the gaps in the development of the IoT services and environments that can be of assistance to the elderly and individuals living with dementia or some sensory impairment. These developments are meant to improve the quality of life for these people, helping with their independence and inclusion, as well as providing assistance and support to their caregivers or families. To see how these outcomes may be achieved, the book outlines the requirements of the systems that aim to furnish some digital sensory or cognitive assistance to the concerned individuals. The book covers the important evolutions of the IoT, smart buildings, cities, and environments. It briefly looks at their enabling technologies, the sensors, actuators, and the communications mechanisms that power up these megatrend infrastructures. The concentration of this work is the use of the IoT-based systems in providing the stated services, improving the conventional AT, and provisions of Ambient Assisted Living (AAL). To elaborate these points, some discussions and samples relevant to how technology in general, and the IoT in particular, can be of value in enhancing human experiences and quality of life are presented. The focus will then be on identifying the missing links, the research gaps, and the challenges in addressing the deficiencies. The book takes an impartial, and yet holistic, view to provide research insights and inspirations for more development works in these areas. It aims to accentuate the need for taking a comprehensive and combinatory view of the comprising topics and approaches, based on the visions and ideas from all stakeholders, particularly the most important ones, the intended users. The book will examine these points and considerations to conclude with recommendations for future works and research directions.

This book can be of value to a diverse array of audience. The intended audience includes the elderly, individuals living with disability, their informal and formal carers. The researchers and developers in healthcare and medicine, aged care and disability services, as well as those working in the IoT-related fields may find many parts of this book useful when aiming to reach out to the experts in the other areas. It is acknowledged that for their own areas of expertise, some professionals may find some parts lacking the depth or the technical content that they may prefer. However, it is worthwhile to mention that, as the discussions in the book will show, one of the major issues that have prevented the development of appropriate and widely-adopted assistive solutions relates to how the medical, healthcare professionals, for instance, view the disruptive technologies, such as the IoT, and their usefulness for the AT developments. Healthcare professionals may hold similar concerns on the views of technologists and perhaps some age and disability advocates, relevant to their understandings of aging, disability, medical, and health-related topics. Either way, a major issue that needs to be addressed is the focus of researchers and developers on their own areas of expertise, without sharing or consulting with professionals from other sectors, even when they are all working on the same types of problems. This book is an attempt in the direction of addressing that kind of issue.

Writing this book took much more work and time than I originally anticipated. It can continue and morph into many bigger projects. As a wise man once told me, you never finish this type of work; you just abandon it. Having said this, I like to thank Mr. Paul Drougas, the Senior Editor of Springer Nature, for all his help and patience. Many people have reviewed this work, from its inception to its final draft. Given the process involved, I do not know the names of most of them. I thank them all, either for their encouraging or constructive comments. In particular, I like to mention Professors Abbas Jamalipour, Friedbert Kohler, and ZhaoYang Dong. I also like to acknowledge and thank Dr. Farnaz Farid and Mr. Ahmed Dawoud for their great help with some of the graphics and data used in this work. Finally, I like to thank my family for their support and understanding while I was working on this book. I sincerely hope this work can be of some value in improving the quality of life for some people.

Sydney, Australia   Seyed Shahrestani

Contents

Abbreviations

AAL	Ambient assisted living
ABS	Australian Bureau of Statistics
AD	Alzheimer's disease
ADA	Americans with Disabilities Act
AHRI	Aware home research initiative
AI	Artificial intelligence
AIOTI	Alliance for Internet of Things Innovation
AP	Access point
APL	Priority assistive products list (WHO)
AT	Assistive technology (and technologies)
BLE	Bluetooth low energy
CRPD	Convention on the rights of persons with disabilities
DDoS	Distributed denial of service
DGPS	Differentially corrected GPS
DSP	Digital signal processing
EC	European Commission
ETA	Electronic travel aids
EU	European Union
FP7	Framework Seven Programme (European Council)
FTC	Federal Trade Commission
GATE	Global cooperation on assistive technology (WHO)
GDP	Gross domestic product
GerAmi	Geriatric ambient intelligence
GPS	Global positioning system
GTSH	Gator Tech smart house
ICF	International Classification of Functioning
ICT	Information and communication technology
IEEE	Institute of Electrical and Electronic Engineers
IFR	International Federation of Robotics
IoT	Internet of things
IR	Infrared

ITU	International Telecommunications Union
LIDAR	Light detection and ranging
M2M	Machinetomachine (communication)
NFC	Near field communications
NSF	National Science Foundation (US)
QoLT	Quality of life technology
RF	Radio-frequency
RFID	Radio-frequency identification
SIG	Special interest group
SLAM	Simultaneous localization and mapping
UGA	Universal gateway for android
UN	United Nations
UWB	Ultra-wideband
WHO	World Health Organization
WLAN	Wireless local area network
WPAN	Wireless personal area networks
WSN	Wireless sensor networks

Chapter 1
Aging, Disability, and Assistive Internet of Things

The World Health Organization (WHO) has estimated that globally more than one billion people, or nearly 15% of the world population, live with disability [1]. It is also estimated that between 2 and 4 percent of the world population experience significant difficulties in functioning. Women, poor people, and the elderly are disproportionately affected by disability. The widely accepted International Classification of Functioning, Disability, and Health considers "disability as an umbrella term for impairments, activity limitations and participation restrictions, denoting the negative aspects of the interaction between an individual (with a health condition) and that individual's contextual (environmental and personal) factors" [2, 3]. As a result of the universal aging of populations, the number and proportion of individuals who experience disability are both increasing. The United Nations (UN) estimates indicate that more than 900 million people, or around 12% of the world population, were over 60 years of age in 2015 [4]. To overcome some of the functional limitations that are associated with age and disability, Assistive Technologies (AT) are employed by many individuals. An overview of age and disability effects along with an outline of how AT can be of value in mitigating some the consequent functional limitations is given in the next section.

The rapid advances in the Information and Communication Technology (ICT) along with the proliferation of the broad range of sensors and actuators that are becoming readily available at low costs have facilitated the explosive growth of the disruptive Internet of Things (IoT) infrastructure. The IoT is drastically reshaping how individuals interact with other people, devices, and their surroundings. It can also furnish some parts of the information that are usually obtained through the human senses, and it is even able to process this data. As such, its deployment for assisting the elderly or individuals with some sensory or cognitive impairment can provide effective and efficient solutions. An overview of the IoT and the smart environments that can be built with it is provided in Sect. 1.2. The discussions on the ways that the IoT and smart environments can complement the more traditional AT are started in Sect. 1.3.

© Springer International Publishing AG 2017
S. Shahrestani, *Internet of Things and Smart Environments*,
DOI 10.1007/978-3-319-60164-9_1

1.1 Aging, Disability, Dementia and Assistive Technologies

Our experiences are heavily reliant on the use of our senses and our cognitive capacities. Unfortunately, a real aspect of life is that both of these can be subject to impairments and deteriorations. The age, illnesses, trauma, accidents, and genetics are among the main reasons for such impairments. When one of our senses, sight, hearing, touch, taste, smell, or spatial awareness is not functioning at "normal" capacity, the condition is considered to be a sensory impairment. However, how some impairment or its underlying causes may relate or lead to disability, have been the subject of different views and interpretations. Section 2.2 provides more discussions on this matter.

The more dominant views of the recent past are the so-called medical and social models [5]. The applied model and the ensuing views have remarkable consequences for the solutions and remedies that are sought and how the individuals are treated. These views and models also influence the design of the related technology and devices, which are commonly referred to as the Assistive Technologies (AT). Hence, and perhaps, more importantly, they indirectly and at least to some degree, determine the adoption and uptake or abandonment of AT by their supposed users.

Irrespective of how the interpretations of the conditions are made, disability appears to be part of the human life. WHO estimates that around one in seven individuals around the world live with some form of disability [6]. Of these one billion people, the *World Health Survey* estimates that 110 million people live with "very significant difficulties in functioning" and the *Global Burden of Disease* assessments put the number of individuals with "severe disability" at 190 million. The severe disability is meant to be equivalent to what is experienced for conditions such as quadriplegia or blindness. The disability experience is, however, more complex than an individual's health conditions. While all individuals, at some stage of their lives, have some form of short-term or permanent impairment, not all are considered to be with disability. The same *WHO Report*, documents pervasive evidence of barriers, other than an individual's impairment, resulting in disadvantages experienced by people with disability.

The disability is related to facing difficulties and complications in functioning. In fact, as further detailed in Sect. 6.4, the very legal definition of disability used in the modern regulatory frameworks and advanced countries, requires that a person's medical condition substantially limit one or more of their major life activities [7]. The use of AT or medications that may mitigate the effects of those limits, by themselves, do not exclude a disability. This statement is related to the fact that these limits can, in reality, be the product of complex interactions of several factors, each complicated in its own right. These factors may include impairment and health conditions, adverse or inadequate regulations and policies, not involving affected individuals and all other stakeholders in decision-making processes, insufficient funding and services, negative personal or social attitudes, accessibility issues, and other barriers. With such complexity, overcoming or even mitigating the disadvantages people with disability face, requires persistent and systemic approaches.

Technology-based solutions can particularly play significant roles for achieving the desired solutions and outcomes.

Numerous forms of AT and Ambient Assisted Living (AAL) services are in widespread use by the elderly and people facing disability for more inclusiveness and independence, as well as for the improved quality of life in other respects [8]. The assistive technologies are considered as items, equipment, or products that can be used to increase, maintain, or improve functional capabilities of individuals with some disability [9]. The AT is a generic term that is used to describe many types of devices and services for people with different impairment types and levels as assistive, adaptive, and rehabilitative tools. These devices and technologies can cover an extensive range of complexity, from, for example, a simple specially designed pen or a walking cane to a smart computer-based communication system [10]. It should be noted that AT always relates to someone's functional capability, rather than to their underlying impairment or age.

Given their broad ranges of applications, different definitions for AT exist. A definition that is widely cited in the literature and is generally accepted by most stakeholders is the one given in the United States, the US Public Law 100-407, the Technical Assistance to the States Act. It defines AT as "any item, piece of equipment or product system whether acquired commercially off the shelf, modified, or customized that is used to increase, maintain or improve functional capabilities of individuals with disabilities" [11]. Their role is to enable people to function in a way that they can participate in everyday life activities. The miscellany of the applications of AT makes it a field that is developing at a fast pace. It encompasses many diverse areas, such as health, design, manufacturing, communication and computer technologies, psychology, and many others. It also covers a vast array of devices and services. These include hearing aids, cochlear implants, eyeglasses and other low vision assistance devices, wheelchairs, walking frames, prostheses and artificial limbs, augmentative and alternative communication approaches such as those that utilize computers, screen-reading software, and customized smartphones [3].

With all their recent advances, AT and AAL still have much room for improvements. A significant aspect of such improvements relates to the well-known and the examined fact that many individuals abandon AT, as the technologies may ignore the social context of their deployments [12]. As pointed out earlier, this has in part been attributed to the different views and the disability models that are based on them. The dominant views, the medical and social models, play central roles in how AT is defined, designed, and is envisioned for use by people with disability or the elderly [13]. The medical model, with origins in biological determinism, has dominated the disability scene for most of the twentieth century [14]. The widespread recognition of this model may have been the consequence of the 1980 WHO classification of terms like "disability" and "impairment" [15]. The "impairment" was defined as "any loss or abnormality of psychological, physical or anatomical structure or function." A "disability" was considered an aftereffect of the impairment, precluding a person from "being able" (hence, disabled) to "perform an activity in the manner or within the range considered normal for a human being" [16]. This model views disability as deviations from some "norm" or some biological function standards. As the model can provide somehow specific technology requirements based on the perceived functional limitations arising from the impairments, it has been the preferred model for many professionals in the last century. As such, it has significantly influenced the AT developments that have been with a naïve or restrictive view of rehabilitation [17].

Many contemporary researchers and most organizations of and for people with disability find the medical model to be oppressive [14, 18]. Their more preferred interpretations are based on the social model. This model sees the physical and social barriers that an affected individual experiences, originating from the society, rather than individual's impairment [19]. In this model, physical, environmental, or social barriers and ignorant attitudes are regarded as the causes of the lost opportunities for those who are deemed to have some disability. This model is now the basis of the more widely accepted view by many professionals, organizations, and governments. WHO has updated its classification and language in light of this model [2]. Based on this view, technology should aim to empower individuals to realize their goals by improving the social structures, removing the barriers, and educating the uninformed persons. This view is used in this work to explore the recent explosive growths of the Internet of Things, smart environments, and their many exciting opportunities for aged care and empowering people with disability or dementia.

1.2   The IoT and Smart Environments

The IoT is drastically reshaping how individuals interact with other people, devices, organizations, their surroundings, and the world. In its contemporary form, IoT is meant to refer to an infrastructure consisting of fairly constantly communicating objects, or "things" that may be smart and process or act on data. Things can be various devices, like medical and health monitoring equipment, medications and drugs, TVs, fridges, and many others, as well as services and software agents. The estimates and forecasts of the numbers of IoT devices have not been easy to make and vary rather significantly from one industry analyst to another [20]. These variations can be seen from a sample of forecasts of connected devices in 2020, depicted in Fig. 1.1. Cisco is among the most cited sources of these estimates and forecasts [22]. It estimates that currently there are around 30 billion IoT connected devices and the number of connected things will rise to 50 billion by 2020 and a staggering 500 billion by 2030. Ericson has forecast that with a compound annual growth rate of 23%, the number of IoT devices will surpass that of mobile phones by 2018 [26].

AT&T and IDC analysis predict that by 2020 the worldwide revenue of IoT will be more than US$7 trillion [24]. IBM puts the economic value of the IoT at US$11 trillion by 2025 and sees it as the largest single source of data globally (http://www. ibm.com/internet-of-things/). Such big data holds key possibilities and values that can be turned into innovations and operational efficiency. The combination of the IoT and big data analytics can be considered as a digital ecosystem.

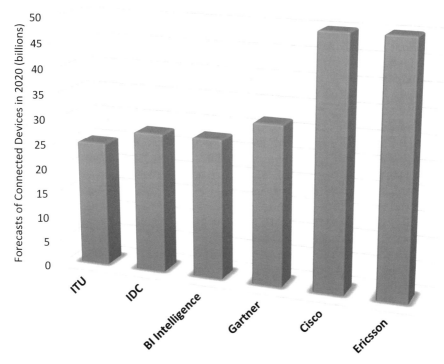

Fig. 1.1 Forecast of IoT Connected Devices in 2020 by Various Industry Analysts. Data sources: [21–25]

There is also a trend for increased dependency of more social and human-based networks, like healthcare, education, power and energy management, financial, or transport on technology-based networks and particularly on the Internet and more specifically on the IoT. The realization of the significant potential advantages of such superimpositions is dependent on being able to analyze and manage the resulting complex systems and also on proper development and deployment of the required applications. Several segments of this work, particularly Sect. 3.3 and Chap. 5, provide many samples of such applications and identify some of the missing parts.

There is no clear and universally accepted definition for the IoT, or for that matter, for things. However, there are at least two main concepts that characterize the IoT. The first one is that most objects or things either have or can have sensing or actuation and communication capabilities. These object may also have other data processing or retention capabilities. The second characteristic is the capacity of direct MachinetoMachine (M2M) communication and control over the Internet. Technically speaking, these mean that all things need to be uniquely identifiable in the global sense of the Internet. Neither of these concepts is that new. The IoT revolution relates to the prospects it provides for building systems using a massive number of interconnected devices that can be quite diverse with a vast range of different capabilities and applications [27].

Section 4.1 provides a more detailed overview of the IoT technologies and related concepts for interactions between humans, digital devices, household equipment, and things in the context of this work. However, given the aims and nature of this book, it cannot be an extensive description. Several good review papers that cover various aspects of the IoT have already been published. The interested reader can, for instance, see [28–32].

The IoT is also the enabler of smart environments and systems, including smart homes, buildings, cities, transport, and healthcare, among many others [33, 34]. The IoT and smart environments can facilitate full and meaningful interactions between humans, digital devices, and many other industrial and household equipment, appliances, and things. They incorporate smart devices and environments with massive interconnections. The pervasive nature of the IoT, comprising of sensors and actuators that are seamlessly interwoven into a physical location, is the foundation of a smart environment. The sensors or actuators can also be smoothly embedded in everyday objects, while connected to other objects and the Internet. The data that the sensors provide can be used for comprehensive monitoring of the environment and surveillance of people or objects within that environment [35].

A Smart City, for instance, has been considered as a transformative approach that uses ICT to improve city administration and infrastructure management. It aims to achieve economic regeneration and better social cohesion, enhancing the residents' quality of life [36]. The goal for a Smart City is to achieve more intelligent use of the physical, cultural, and public resources to increase the quality of the services offered to the citizens, more efficiently with lower costs [37]. The Fourth Version of the European model considers an urban area as a Smart City if it performs well in the six identified key development fields "built on the 'smart' combination of endowments and activities of self-decisive, independent and aware citizens" (http://www.smart-cities.eu). The key fields are the smart economy, smart governance, smart living, smart mobility, smart environment, and smart people. Smart buildings, cities, and environments and the new solutions that they can provide for disability, dementia, and aged care are further discussed in Sects. 4.2 and 5.1.

1.3 The IoT, Assistive Devices, and Technologies

The extensive range of available sensors can furnish some parts of the information usually obtained through the human senses, for instance through eyesight. While the level of sophistication of sensors is increasing, their prices are rapidly falling [38]. Combining these advances with the diverse range of communication technologies that include Bluetooth Low Energy (BLE) [39], 6LowPAN [40], and Near Field Communications (NFC) (http://nearfieldcommunication.org/), which are suitable for their data exchanges, facilitate what may be referred to as the digital senses. As it will be discussed in Sect. 3.3, such digital senses, augmented with the permeating nature of the IoT and deployment of smart environments, can be of significant benefit to an elderly or a person with some sensory of cognitive impairment.

Smartphones can also play a leading role in various implementations that are based on the IoT and smart environments. While they carry an increasing number of sensors, they can also facilitate most of the communications between the IoT devices and the non-local, external, or cloud-based services. The smartphones can also do some of the required processing and act as the interface for interactions between the machines or things and people.

With their prevalence and diverse capabilities, IoT, smart homes, smart buildings, smart cities, and other smart environments can overcome some of the challenges that the more traditional AT face. They have the potential to enhance human experiences and quality of life extremely. More specifically, with their potentially ubiquitous nature and large growths, such technologies and infrastructure can and should play leading roles in the aged care and improve the lives of people with disability or dementia. However, this does not appear to be the case in practice. The issues relevant to the proliferation of the assistive IoT and smart environments will be discussed throughout this work, particularly in Sect. 5.2 and Chap. 6. Some samples of the deployments of these technologies are presented in Sect. 5.1.

While the IoT-based AT has great possibilities, there is still much room for work and improvements in these areas, particularly regarding their applications, implementations, and usability or adoption. A general aspect that needs particular attention relates to the requirement for taking a holistic approach that can provide or assist with the enhanced independence, access provisions, and integration of the affected people while providing support for their families and caregivers. These are the areas that innovations and research are truly required. With the aim of identifying the issues that prevent the widespread adoption of IoT-based solutions and to motivate research and implementations that can overcome them, some of the challenges in moving forward are considered in Sect. 5.2.

There are many interesting available or in-progress works on the use of assistive IoT and smart technologies. This book, aware of its space limitations and the patience of its potentially diverse range of readers, looks at a cross-section of those works. The book also takes an impartial, and yet holistic, view to provide research insights and inspirations for more development work in these areas. It aims to accentuate the need for taking a comprehensive and combinatory view of the comprising topics and approaches, based on the visions and ideas from all stakeholders, particularly the most important ones, the intended users.

References

1. World Health Organization. (2011). *World report on disability*. World Health Organization. Also available online http://apps.who.int/iris/bitstream/10665/70670/1/WHO_NMH_VIP_11.01_eng.pdf.
2. World Health Organization. (2001). *International Classification of Functioning, Disability and Health: ICF:* World Health Organization.

3. World Health Organization. (2015). *WHO global disability action plan 2014–2021: Better health for all people with disability*. World Health Organization.

4. United Nations, Department of Economic and Social Affairs, Population Division (2015). World Population Ageing 2015 (ST/ESA/SER.A/390). Available online, http://www.un.org/en/development/desa/population/publications/pdf/ageing/WPA2015_Report.pdf   [Accessed 24th March 2017].

5. Benjamin Darling, R., & Alex Heckert, D. (2010). Activism, models, identities, and opportunities: A preliminary test of a typology of disability orientations. In Disability as a Fluid State: Research in Social Science and Disability (pp. 203–229). Emerald Group Publishing Limited.

6. World Health Organization. (2011). *World report on disability*. World Health Organization.

7. Meeks, L. M., & Jain, N. R. (2015). *The Guide to Assisting Students With Disabilities: Equal Access in Health Science and Professional Education*. Springer Publishing Company.

8. Islam, S. R., Kwak, D., Kabir, M. H., Hossain, M., & Kwak, K. S. (2015). The internet of things for health care: a comprehensive survey. IEEE Access, 3, 678–708.

9. Cook, A. M., & Polgar, J. M. (2014). *Assistive technologies: Principles and practice*. Elsevier Health Sciences.

10. Maor, D., Currie, J., & Drewry, R. (2011). The effectiveness of assistive technologies for children with special needs: a review of research-based studies. *European Journal of Special Needs Education, 26*(3), 283–298.

11. Wendt, O., & Lloyd, L. L. (2011). Definitions, history, and legal aspects of assistive technology. Assistive technology: Principles and applications or communication disorders and special education, 1–22.

12. Shinohara, K., & Wobbrock, J. O. (2011, May). In the shadow of misperception: assistive technology use and social interactions. In *Proceedings of the SIGCHI Conference on Human Factors in Computing Systems* (pp. 705–714).

13. Hersh, M. A., & Johnson, M. A. (2008). Disability and assistive technology systems. In *Assistive technology for visually impaired and blind people* (pp. 1–50). Springer London.

14. Frauenberger, C. (2015). Disability and technology: A critical realist perspective. In *Proceedings of the 17th International ACM SIGACCESS Conference on Computers & Accessibility* (pp. 89–96).

15. World Health Organization. (1980). International classification of impairments, disabilities, and handicaps: a manual of classification relating to the consequences of disease; publ. for trial purposes in accordance with resolution WHA29. 35 for the Twenty-ninth World Health Assembly, May 1976.

16. Hersh, M. A., & Johnson, M. A. (2008). *Assistive technology for visually impaired and blind people*, 167–208. Springer London.

17. Mankoff, J., Hayes, G. R., & Kasnitz, D. (2010, October). Disability studies as a source of critical inquiry for the field of assistive technology. In *Proceedings of the 12th international ACM SIGACCESS conference on Computers and accessibility* (pp. 3–10).

18. Shakespeare, T. (2013). *Disability rights and wrongs revisited*. Routledge.

19. Swain, J., French, S., & Cameron, C. (2003). *Controversial issues in a disabling society*. McGraw-Hill Education (UK).

20. Nordrum, A. (2016). Popular internet of things forecast of 50 billion devices by 2020 is outdated. IEEE Spectrum, 18.

21. BI Intelligence. (2015). The Internet of Everything. Available online., http://www.businessinsider.com/iot-ecosystem-internet-of-thingsforecasts-and-business-opportunities-2016-2/?r=AU&IR=T [Accessed: 10-Apr-2017].

22. Evans, D. (2011). The internet of things: How the next evolution of the internet is changing everything. *CISCO white paper, 1*(2011), 1–11.

23. Lueth, K. (2014). IoT Market – Forecasts at a glance. Available online, https://iot-analytics.com/iot-market-forecasts-overview/ [Accessed 24th March 2017].

24. Lund, D., MacGillivray, C., Turner, V., & Morales, M. (2014). Worldwide and regional internet of things (iot) 2014–2020 forecast: A virtuous circle of proven value and demand. *International Data Corporation (IDC), Tech. Rep.*

25. Walport, M. (2014). The Internet of Things: making the most of the Second Digital Revolution. *London:* UK Government Office for Science. Available online. https://www.gov.uk/government/uploads/system/uploads/attachment_data/file/409774/14-1230-internet-of-things-review.pdf

26. Cerwall, P., Jonsson, P., Möller, R., Bävertoft, S., Carson, S., Godor, I., & Lindberg, P. (2015). Ericsson mobility report. *Ericson, Tech. Rep.* Available online: https://www.ericsson.com/res/docs/2015/mobility-report/ericsson-mobility-report-nov-2015.pdf [Accessed 24th April 2017].

27. Greengard, S. (2015). *The Internet of things.* MIT Press.

28. Atzori, L., Iera, A., & Morabito, G. (2010). The internet of things: A survey. *Computer networks, 54*(15), 2787–2805.

29. Li, S., Da Xu, L., & Zhao, S. (2015). The internet of things: a survey. *Information Systems Frontiers, 17*(2), 243–259.

30. Miraz, M. H., Ali, M., Excell, P. S., & Picking, R. (2015, September). A review on Internet of Things (IoT), Internet of Everything (IoE) and Internet of Nano Things (IoNT). In *Internet Technologies and Applications (ITA), 2015* (pp. 219–224).

31. Stojkoska, B. L. R., & Trivodaliev, K. V. (2017). A review of Internet of Things for smart home: Challenges and solutions. Journal of Cleaner Production, 140, 1454–1464.

32. Yang, D. L., Liu, F., & Liang, Y. D. (2010). A survey of the internet of things. In *Proceedings of the 1st International Conference on E-Business Intelligence (ICEBI2010).* Atlantis Press.

33. Elkhodr, M., Shahrestani, S., & Cheung, H. (2015). A Smart Home Application Based on the Internet of Things Management Platform. In Data Science and Data Intensive Systems (DSDIS), 2015 IEEE International Conference on (pp. 491–496).

34. Pellicer, S., Santa, G., Bleda, A., Maestre, R., Jara, A., & Gomez Skarmeta, A. (2013). A Global Perspective of Smart Cities: A Survey. *Innovative Mobile and Internet Services in Ubiquitous Computing (IMIS), 2013 Seventh International Conference on, 439–444.*

35. Fernández-Caballero, A., Martínez-Rodrigo, A., Pastor, J. M., Castillo, J. C., Lozano-Monasor, E., López, M. T., & Fernández-Sotos, A. (2016). Smart environment architecture for emotion detection and regulation. *Journal of Biomedical Informatics, 64,* 55–73.

36. Ojo, A., Curry, E., & Janowski, T. (2014). Designing next generation smart city initiatives-harnessing findings and lessons from a study of ten smart city programs. In *Proceedings 22nd European Conference on Information Systems.*

37. Zanella, A., Bui, N., Castellani, A., Vangelista, L., & Zorzi, M. (2014). Internet of things for smart cities. *IEEE Internet of Things journal, 1*(1), 22–32.

38. Meunier, F., Wood, A., Weiss, K., Huberty, K., Flannery, S., Moore, J., & Lu, B. (2014). *The Internet of Things Is Now: Connecting the Real Economy.* Technical Report.

39. Bluetooth SIG, Bluetooth Low Energy. (2016). .Available online, https://www.bluetooth.com/what-is-bluetooth-technology/bluetoothtechnology-basics/low-energy [Accessed: 10-Oct-2016].

40. Shelby, Z., & Bormann, C. (2011). *6LoWPAN: The wireless embedded Internet* (Vol. 43). John Wiley & Sons.

Chapter 2
Assistive IoT: Enhancing Human Experiences

Our perception of the world and what goes on around us is a function of obtaining information through our senses, which is then processed and interpreted through the use of our cognitive abilities and memory. Mobility is also dependent on perception, along with the motor and other skills and abilities. While we heavily rely on our sensory and cognitive abilities, their imperfections and impairments are part of the human life. They can occur due to genetic reasons, illness, trauma, injury, or age. Nevertheless, sensory or cognitive impairments can affect human experiences and quality of life. For instance, the ability to move about or travel independently is a human desire. This desire holds for all individuals, including those who are blind or visually impaired. For many seniors, particularly for those living with dementia, memory loss or decline can be a highly distressing experience. Furthermore, many people with sensory, cognitive, or motor impairments often encounter numerous barriers in their daily life. In this chapter, after considering the scale and the effects of sensory impairments, dementia, and aging, the main views on disability will be presented. It will then contemplate on the AT requirements and the disability and aging experiences. These lead to detailed discussions on how technology in general, and the IoT in particular, can be of value in enhancing human experiences and quality of life.

2.1 Disability, Aging, and Dementia: Scales and Impacts

With their scale and numbers, people affected by disability, dementia, aging, or a combination of these can be considered as the world's largest minority. To make any meaningful progress, the effects of these conditions on the individuals, their families and caregivers, social infrastructure, economy, among many other areas need to be understood. The next section provides an overview of the scale, growth rates, and

© Springer International Publishing AG 2017
S. Shahrestani, *Internet of Things and Smart Environments*,
DOI 10.1007/978-3-319-60164-9_2

demography of disability, dementia, and aging. Their individual and societal effects, along with impacts on family members and caregivers, as well as on community, economy, and expenditure are discussed in Sect. 2.1.2.

2.1.1   Disability, Aging, and Dementia: Demography and Scale

The number of individuals living with some sensory or cognitive impairment or assisting an affected person is enormous. The estimates for adults living with some form of disability varies between one in seven to one in five people, or between more than 15 to 19 percent of the world population who are over 15 years of age [1]. These estimates for children with disability range from 5% to 8%. Overall, it is estimated that more than one billion individuals live with disability. A snapshot of disability and some of its effects is provided in Fig. 2.1.

The available evidence indicates that sight and hearing impairments are the two most commonly encountered sensory losses. WHO puts the worldwide number of individuals with a disabling hearing loss at 360 million, or around 5% of the world's population [2]. Based on WHO estimates, worldwide, 285 million people are visually impaired, with 39 million blind people, and 246 million individuals have low vision. Some 90% of the blind people are estimated to live in the developing countries [3]. With no comprehensive registry and various parameters considered for characterization of impairments, such as different definitions or age ranges, it can be difficult to get an overall prevalence rate of visual impairments and blindness even for countries like the US [4]. However, using the results of the 2012 National Health Interview Survey Preliminary Report, it is estimated that 20.6 million adult Americans, or around 10% of adult Americans, are either blind or "have trouble" seeing, even when using corrective lenses [5].

The Australian Bureau of Statistics (ABS) reports that in 2015 over 18% of the nation population, or 4.3 million individuals, report disability [6]. It is estimated that one in six Australians are affected by hearing loss and around 30,000 individuals with total hearing loss use Deaf Auslan. Vision Australia puts the approximate number of Australians who are blind or have low vision at 357,000 people (http://www.and.org.au/pages/disability-statistics.html).

The aging population also has great human, economic, social, and health effects for the world. Aging and drastic increases in the number of those living to older ages, compared to previous generations, is a fact of life for many parts of the world. However, the aging is by no means a uniform phenomenon across the globe. Figures from the Department of Economic and Social Affairs of the United Nations, indicate that 901 million people, or around 12% of the world population of 7.3 billion, were over 60 in 2015 [7]. The number of older individuals is projected to reach 1.4 billion by 2030 and 2.1 billion by 2050. The number of people over 60 is the fastest growing slice of the global population with an increase-rate of 3.26% per annum, much higher than the overall population growth rate of 1.18% per year.

15%
of the World Population Lives with Some Sort of Disability

80% of people with disabilities live in developing countries.

STATISTICS OF COMPLETING PRIMARY SCHOOL FOR MALES

50.6% MALES WITH DISABILITY

61.3% MALES WITHOUT DISABILITY

STATISTICS OF COMPLETING PRIMARY SCHOOL FOR FEMALES

52.9% FEMALES WITHOUT DISABILITY

41.7% FEMALES WITH DISABILITY

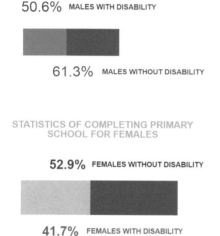

The employment rates are lower for disabled men (53%) and disabled women (20%)than for non-disabled men (65%) and women (30%).

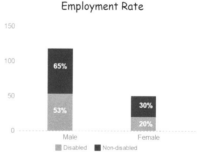

Fig. 2.1 Snapshot of the world disability. Various data sources cited in the text

While the aging is a global concern, its scale and its consequences vary significantly amongst different parts of the world. Some of the differences are depicted in Figs. 2.2, 2.3, 2.4, and 2.5. Sixty-six percent of the aging population increase, between 2015 and 2050, is estimated to occur in Asia. For this continent, it is expected that the proportion of people aged 60 or over to more than double between 2015 and 2050, from 12% to 25% of the population. For Northern America, these changes are from 21% in 2015 to 28% by 2050, and for Oceania from 16% to 23%. It is projected that globally, the number of persons aged 80 or over will increase from 125 million in 2015 to 434 million in 2050 [9]. That corresponds to a 250% increase of people in this age bracket in 35 years.

The aging has clear influences on the disability prevalence, as the potential of disability increases with older age. While the prevalence of moderate and severe disability for people of all ages is around 15.1%, the figure for those over the age of

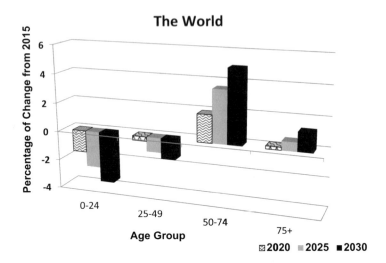

Fig. 2.2 Aging effect: the world (Percentages of changes in age groups from 2015). Data Source: [8]

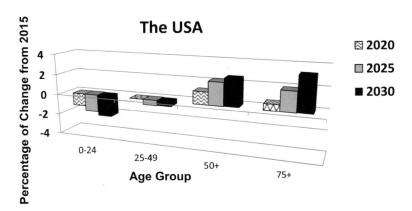

Fig. 2.3 Aging effect: the USA (Percentages of changes in age groups from 2015). Data Source: [8]

60 years is estimated to be 46.1% [1]. The buildup and increased risks of trauma, injuries, and diseases over a long lifespan are considered to attribute to the higher disability rates among older people. It has been reported that almost half of all men and around one-third of all women over the age of 65 are affected by hearing loss [10]. Vision is also commonly affected by age. It is estimated that in the Western countries, the proportion of people who are visually impaired or blind increases by a factor of three per decade after the age of 40 [11]. The frequency of cataracts, the cause of 45% of blindness instances globally, increases with age, having a prevalence rate of 40% for people over the age of 70 years [3].

According to 2015 ABS reports, in Australia, people over 65 years old constitute 15.1% of the population, increased from 14.3% in 2012. The data shows that 50.7%

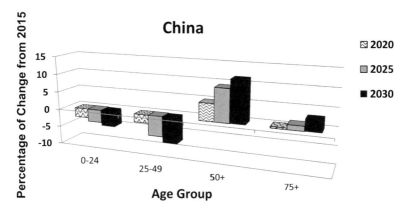

Fig. 2.4   Aging effect: the world (Percentages of changes in age groups from 2015). Data Source: [8]

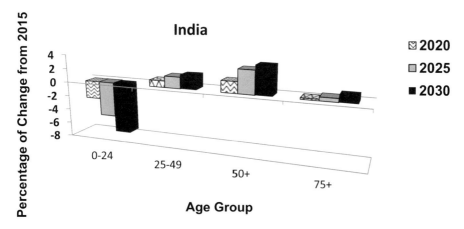

Fig. 2.5   Aging effect: India (Percentages of changes in age groups from 2015). Data Source: [8]

of these individuals were living with disability, although decreased from 52.7% in 2012 [6]. The proportion of Australian people over 55 years old who live with some disability is reported to be 43% of the population of this nation. Within this age bracket, again, being older increases the chance of disability. That is, 31% of 55 to 64 year-olds are living with disability, whereas around 88% of those aged over 90 have a disability (http://www.and.org.au/pages/disability-statistics).

Cognitive abilities vary widely among people, but they may also correlate with age. While some deteriorations of these abilities naturally occur with aging, some disorders may have more effects on them. Dementia is an example of such disorders. Dementia ensues from an acquired brain ailment, resulting in progressive deterioration of cognitive functions. It can affect memory, orientation, learning capability, reasoning, judgment, emotional control, social behavior, and motivation [12].

The final stages of dementia are usually accompanied by loss of memory and reasoning capabilities, in addition to the deterioration of speech and other cognitive-related functions. Based on WHO studies, in 2015 more than 47 million people worldwide have been affected by dementia [13]. The number of people living with dementia is estimated almost to double every 20 years, getting close to 75 million by 2030 and increase to 131.5 million by 2050 [14]. Alzheimer's Disease (AD) is the most common form of dementia, accounting for 60–80% of all cases [15]. Alzheimer's Disease International estimates that globally, near ten million new cases of dementia are diagnosed each year. They have also concluded that "the incidence of dementia doubles with every 6.3-year increase in age, from 3.9 per 1000 person years at age 60–64 to 104.8 per 1000 person years at age 90+" [14]. This trend is witnessed in other studies from different corners of the world. For example, in the United States, more than 95% of the 5.4 million people diagnosed with AD in 2003, were over 65 years of age. That corresponded to one in eight people over 65 having this disease [4]. In Japan, around 10% of the elderly population, corresponding to about 2.8 million individuals, live with severe dementia. A further 3.5–5 million people are estimated to live with moderate dementia [16]. A summary of the scale of dementia and its growth, worldwide, is shown in Fig. 2.6. The global numbers of individuals living with some sensory impairment or dementia are summarized in Fig. 2.7.

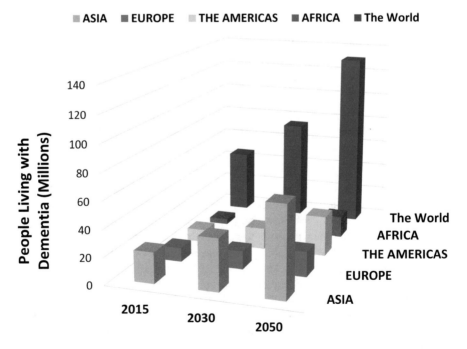

Fig. 2.6 The Number of people Living with Dementia. Data Source: [14]

Fig. 2.7 Sensory and cognitive impairments in the world

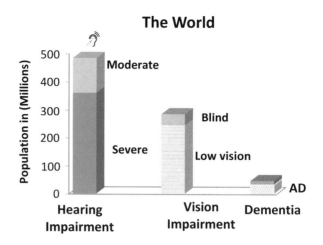

2.1.2 Disability, Aging, and Dementia: Effects and Impacts

With life expectancies of over 70 years in many parts of the globe, on average, a person with disability lives for about 8 years, or 11.5% of their lifetime [17]. A significant majority of around 80% of people with disability live in developing countries [18]. Poverty and disability can form a vicious cycle. Through unemployment or underemployment, lower education prospects, lower income, and escalated living or medical expenses disability can increase poverty risk factors. Poverty usually goes with malnourishment, poor access to quality education and healthcare, hazardous working environments, and limited access to clean water and sanitation, all of which can again, increase the risk of disability. So, it is not surprising that The World Bank estimates that 20% of the world's poorest people are those living with a disability (http://www.worldbank.org/en/topic/disability/overview).

For individuals with disability, as for all people, employment is not only important for having an income to pay for living expense, but it can also help with social integration, self-respect, and a sense of drive for daily life. Based on analysis of the World Health Survey results for 51 countries, the WHO estimates the employment rates of people with disability to be as 52.8% for men and a mere 19.6% for women, while for the general population the rates are 64.9% and 29.9% respectively [1]. A summary comparison of these employment rates appears in Fig. 2.8. In some countries, people with disability face 80% unemployment rates. In the United States the results of a 2004 survey showed that while 78% of the general working-age population are actually working, only 35% of individuals with disability enjoy the same condition [17].

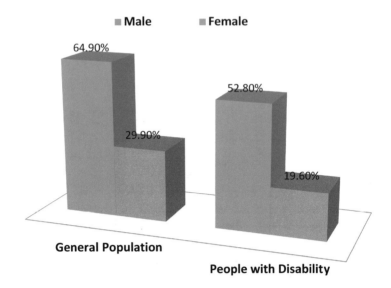

Fig. 2.8 Comparison of Global Employment Rates. Data Source: [1]

Table 2.1 Employment Rate of People with Disability in some developed countries.

Australia	47.7%
Canada	49%
Denmark	43.9%
Luxembourg	48%
New Zealand	45%
Norway	43%
United Kingdom	48.9%

Data Source: [1]

High unemployment rates can be the consequences of negative social attitudes and stigma that can also lead to low self-esteem and reduced participation for some of the people living with disability or the elderly. For this or other reasons, some individuals may not fully appreciate their abilities for employment and not even participate in the labor force. In Australia, for instance, with a general participation rate of those aged 15–64 years being 82%, only 53% of people with disability in the same age bracket participate in the labor force [6]. The unemployment rates of these working-age individuals at 9.4% is significantly higher than 4.9% of the general Australian population. As Table 2.1 indicates, the 'employment' rates of people with disability in most of the developed countries are comparable but much lower than that of the general population [1].

Table 2.1 may not even provide the right picture for all conditions. For instance, Table 2.2 shows a drastically different situation with the 'unemployment' rates of the working age blind and visually impaired people in some European Union (EU)

Table 2.2 Unemployment rate of the blind or visually impaired people in some EU countries

Finland	55%
Germany	72%
Hungary	77%
Norway	68%
Poland	87%
Spain	4.2%
Sweden	5.5%

Data Source: [19]

countries, based on a 2001 survey [19]. It has also been pointed out that "… in Spain 85% of the members of the national blind association who are in employment, sell Lottery of the Blind tickets. Clearly, this type of employment does not have a career structure with promotion prospects and it is unlikely that it is matched to the skills and interests of 85% of blind people..." [19].

More generally, for people with disability, underemployment with all of its conditions, can be another major issue, even when they are employed. For example, in the United Kingdom, less than half of the nation's 6.9 million people of working age with disability have jobs (http://www.dlf.org.uk/content/key-facts). A 2014 analysis found that those working "were more than twice as likely as peers without disabilities to report working part-time, and about half as likely to have jobs in science, technology, engineering and mathematics". Similar results were reported by the US National Science Foundation in 2015, reporting that around 11% of the scientists aged 75 or younger had a disability and that "they, too, were more than twice as likely to be out of the labor force than their peers without disabilities" [20].

In many situations, isolation and not accessing the social networks of family, friends, and others reduce the chances of finding employment for people with disability or the elderly, which can lead to lower participation in the work force. However, while most jobs can be done by an elderly person or someone with a disability given the right setting, some employers continue to see these individuals as a burden or not productive enough. Fortunately, these views have changed in many corners. Many companies report that the elderly or people with disability are often loyal, hard-working, with low rates of absenteeism, and with the right skill sets [1]. People with disability have also proved to be successful in running their small businesses. One indication of that is provided by the 1990 census in the United States. It showed that 12.2% of people with disability have self-employment and small business experience, which is at a higher rate than that of the general population at 7.8% [17].

The employment rate for all people improves with education, higher qualifications, and better skills. The employment rate for persons with disability, who have completed high school is estimated to be around 30%. Obtaining post-secondary degrees increase their employment rates by about 20% to near 50% [4]. However, significant accessibility barriers have resulted in lower participation by people with disability in education and significantly lower rates of obtaining qualifications

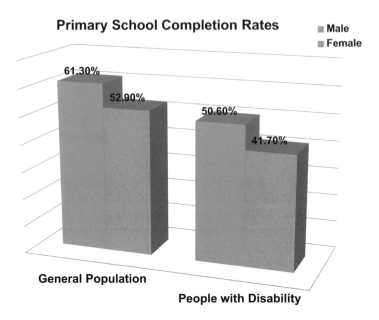

Fig. 2.9 Comparison of Primary School Completion Rates. Data Source: [1]

compared to the population as a whole. For both children and adults, low educational outcomes correlate strongly with disability, even when compared to more publicized parameters including gender, rural residence, and low economic status [1]. Primary school completion rates, as indicative educational outcomes, are compared in Fig. 2.9, using WHO data [1].

As with employment, educational enrollments and completion rates, depend on the impairment type and conditions. For instance, studies in the US have shown that educational attainment odds for people with hearing loss were 3.2 times lower than other individuals. In 1994/95, only 45% of blind or severely visually impaired individuals living in the US had completed high school, which is much lower than 80% of sighted people. In the UK in 1994/95, 0.13% of undergraduate students and .06% of postgraduate students were blind or visually impaired. While 10 years after that, these figures had risen to 0.16% and 0.12% respectively, the proportions are still well below their corresponding estimates of 0.8–2.0% for the population as a whole [19].

Many of the secondary outcomes in employment and education are the consequence of barriers and discrimination faced by people with disability. For instance, 8.6% of Australians with disability have reported the experience of discrimination because of their disability. Notably, more than 35% of women and 28% of men 15 years or older have avoided situations because of their disability [6]. The barriers include inaccessible buildings, lack of appropriate means of transport, extra and added medical or regular living and maintenance costs, and lower access to the ICT (http://www.worldbank.org/en/topic/disability/overview). Access to information and ICT is markedly important. As access to most of the information and ICT interfaces depend on the auditory and visual senses, people with a sensory impairment may again be disadvantaged. As it will be discussed in Sects. 2.4, 3.1, 5.1, and 5.2, these areas have witnessed significant progress in the recent past, but even so, they lack the required coordination between the many moving parts.

While aging can increase the risks and severity of sensory and cognitive impairments and what goes with that, investigations and understanding of aging characteristics, effects and impacts are significant in their own rights. For instance, studies on how demographic changes affect economic potentials are relatively restricted to the outlining of the needs and special demands of the elderly. With the growing proportion of the elderly relative to the whole population, and the corresponding increased market size and power, companies and organizations can only succeed if their products and services are "good for all," independent of the user's age and [21].

Most seniors live in households, rather than in nursing homes or cared accommodations. In Australia, for instance, almost 95% of the elderly live in households [6]. This status is, generally speaking, irrespective of their abilities and includes most people with dementia who are commonly cared for by their informal caregivers, their spouse or relatives [22]. It is estimated that around 30% of the elderly with dementia live alone [23].

Dementia can be overwhelming for the families and the caregivers, as well as for the affected individuals. The Alzheimer's Association points to the fact that in 2015, just for the US, around 16 million family members and friends provided 18.1 billion hours of unpaid care to those with AD and other dementias [24]. The AD International estimates that in 2015, the global cost of dementia has been US$818 billion, which represents 1.09% of global Gross Domestic Product (GDP). It becomes a trillion-dollar disease by 2018. "With no known cure on the horizon, and with a global aging population," they call on "…every part of society to play an active role in helping to create a world where people can enjoy a better quality of life..." [25]. Between 2010 and 2015, the costs of dementia have increased by 35.4% [14]. In the US, the yearly costs of caring for people with AD and other dementia were estimated to have been US$203 billion in 2013 and are projected to reach US$1.2 trillion by 2050 [26]. For the 27 countries of the EU, the costs of long-term care for AD and dementia is forecast to increase from 1.2% of the GDP to 2.5% of it [12].

The three major cost categories for dementia and their relative proportions are shown in Fig. 2.10, using the data from the World Alzheimer's Report [14]. It should be noted that in general, informal care is significantly less costly than accommodations in a nursing home or similar care facilities [12]. The informal carers provide significant economic and social benefits to the society. As it will be further discussed in Sects. 2.3 and 6.3, taking a holistic view of AT needs their input, while the development and implementation of technology that can support them are also a requirement. With increases in the number of people needing care, obviously, the number of carers needs to grow too. However, the very demographic changes themselves will result in a decrease in the number of potential informal caregivers over time [12]. Nevertheless, the scale of informal care is large. For example, in 2015 in Australia, 11.6% of the population were carers, with a massive 3.7% of those over the age of 15 being primary carers. These numbers have not changed significantly

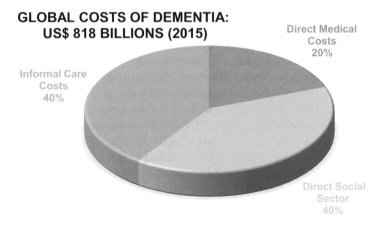

Fig. 2.10 Components of the Global Costs of Dementia in 2015. Data Source: [14]

since 2009. With an average age of 55 years, nearly 38% of the primary carers, were living with disability themselves. Their labor force participation of 56% was significantly lower than the 80.3% for non-carers [6].

2.2 Dominant Disability and Aging Models

As with any other systems or complex conditions, models can prove to be useful for understanding and explaining the underlying concepts and issues in disability studies and aged care. Formation of policies and strategies by governments and societies can also benefit from properly devised and developed models. Similar to most other social and human-related topics, the views of experts and specialists can substantially influence these models, while they actually are for and about other people. As such, while these models can provide some expert insight, they are not complete or necessarily free from prejudices and attitudes of specialists towards the remedial steps for some impairment or perceptions about age. The experts may not be aware of the impact of the impairment, age, or the functional limitations on an affected individual. Given these, many people, treat many of these models with skepticism, considering them not reflecting the real world issues, often encouraging narrow thinking or views, and rarely capable of providing detailed plans for positive actions [27]. The disability models can be limiting or empowering access and inclusion, or at least they can reveal the intentions of those who develop or use them. So, they can facilitate communication and enable various stakeholders to reach some common understanding.

The disability models have been devised to characterize some impairment and how it may affect equal access and participation and the difficulties that people may face in a functioning area. WHO, for instance, acknowledges that "Disability is the umbrella term for impairments, activity limitations, and participation restrictions, referring to the adverse aspects of the interaction between an individual and that individual's contextual factors, including environmental and personal factors." The Preamble to the CRPD acknowledges that disability is "an evolving concept," but also stresses that "disability results from the interaction between persons with impairments and attitudinal and environmental barriers that hinder their full and effective participation in society on an equal basis with others" [1].

Two fundamental philosophies have influenced the more recent disability models and views. The first view looks at the impairment or some medical condition as the cause of disability. The affected individual is therefore seen as dependent upon the society. This view can lead to paternalism, exclusion, and discrimination. The second philosophy sees the disability as a perceived concept arising from barriers, attitudes, and stigma. The individual with an impairment is considered as another member of the society whose constituents have diverse abilities. This view results in sociable choices, inclusion, social empowerment, and equality. The effects of each view and the extent of their applications will be discussed shortly. The models have developed over time, with improvements being made as the result of ongoing changes in social attitudes to disability. Models have changed as society has progressed. With this in mind, the current thinking is to develop and work with models that empower people with disability and facilitate their full and equal participation in social functions and activities [27].

The major WHO Reports have largely influenced the most widely accepted contemporary models. The 2011 Report on disability, for instance, uses three key concepts of impairment, limitation, and participation to characterize disability. It states, "Disability refers to difficulties encountered in any or all three areas of functioning... "Health conditions" are diseases, injuries, and disorders, while "impairments" are specific decrements in body functions and structures, often identified as symptoms or signs of health conditions. Disability arises from the interaction of health conditions with contextual factors – environmental and personal factors ...." [1]. This view is based on the International Classification of Functioning (ICF), Disability and Health [28]. ICF has been instrumental in understanding disability and its impacts. The major advancement of ICF relates to highlighting environmental elements as the major contributor to the disability experience.

The ICF and the views based on it are not that far from the expressions made by Disabled People's International (http://www.dpi.org/). In 1981, they considered disability and impairment to mean two different concepts; "Impairment is the loss or limitation of physical, mental or sensory function on a long-term or permanent basis. Disablement is the loss or limitation of opportunities to take part in the usual life of the community on an equal level with others due to physical and social barriers." [29].

The dominant view about disability in the last century was the "medical model." Disability was primarily a medical term that came from biological determinism [30]. The underlying reason for disability in this model is the individual. Consequently, for their advancement, or to minimize their burden on others, segregation, for example through special needs schools, was justified. Disability was essentially a concept based on deviations from some "norm" for a biological function, where the society's response must be to "cure," "care," or, at best, "support" the affected individual [31].

The effects of social attitudes and physical barriers in creating disability have been long established [1]. A wheelchair user who has a mobility impairment can be mentioned as an example. In an environment where they have full access to buildings and can use public services, like public transport, functionally similar to someone without their impairment, the disability is not of concern [27]. Taking these into account, a transformation of the view of disability towards a "social model" has occurred. This model is meant to empower people by removing the barriers so that all individuals have the same opportunity, although some may be using some form of AT. This contemporary model of disability is widely accepted and incorporates the United Nations Convention on the Rights of Persons with Disabilities (CRPD) [32]. In this model, disability is defined as a category of difference, questioning the concepts arising from "being normal" [31]. In this fashion, people with disability, are similar to other non-mainstream individuals, members of minority groups, who are possibly disadvantaged because of their race, gender, country of origin, or sexual orientations. With over one billion members, people with disability are the world's largest minority group.

The social model has also been referred to as the "Minority-Group Model of Disability" [27]. Such a reference is meant to emphasize that disability stems from the failure of society and its political leadership in meeting the requirements and objectives of a large minority, in some sense similar to say, racial discrimination or equality. So, following on from the wheelchair user example, if they are not able to access a building or use a bus, the building paths or the bus need to be redesigned, or the wheelchair must be modified. Either way, it is not the individual's fault or responsibility. However, this view does not ignore the many aspects that make up the disability experience. The view notes that most of the human experiences are shaped by a combination of factors inherent from the individual and those influenced by external or structural components. Disability is not and should not be an exception to that.

Individual or inherent factors can include the severity and the characteristics of the impairment, and personality or a person's own attitudes towards it. External and structural components can be, for example, the attitudes of others, the physical barriers, improper sound systems, issues with transport systems, socio-economic problems, or environment [31]. This view has been related to the ICF and forming a holistic approach, looking at disability causes coming from society and individual bodies [33]. However, the subjectivity of the influences of each cause and the extent of the interaction of the two causes can make this view, subject to broad interpretations [34]. Nevertheless, with some added objectivity, considering the individual and external factors can facilitate identifying practical ways to improve the quality of life for people with disability. For example, encouraging an individual to employ their AT and coaching them to use it correctly, while externally enforcing appropriate regulations and legislations, are both required. The interplay of these two causes may prove to be difficult to assess objectively. For identifying productive and pragmatic ways to move forward, more works in these areas are clearly needed. As such, some people with disability as well as some researchers in the related areas see the need for moving beyond this somehow simplified standing to augment the model with more complex, but objective, measures.

There are also two main views on aging [35]. The concentration of one of the two views is on improving the quality of life for the elderly through supporting their physical health. The physical healthiness is seen as the means of reducing medical expenses and mitigating the burden of care on the society. The focus here is on physical activities and devices or services that promote health. The second view concentrates on the motivations and the will to live well, irrespective of age and physical well-being. This philosophy is based on accepting the age, physical conditions, disability, ailment, or demise and using medicine, AT, or other means to reduce their negative effects to improve the individual's quality of life. These models are not as mature as those for disability. The lack of well-developed and widely accepted models is also the case for dementia care. While there have been some attempts to develop social disability models for dementia care, they have not been widely applied or accepted for cognitive impairment [36].

Whether these views are related to disability or aging, they can shape the approaches that are taken towards the design of AT and how the affected individual and society perceive the AT. While the social model of disability is recognized to be appropriate in the context of activism and engagement, it has been criticized due to failing to provide tangible steps and requirements for action to support the elderly or people with disability [31]. With that consideration, some also argue that removal of barriers and high levels of access to technology will not necessarily amount to complete integration of affected individuals in a digital society. Rather, it is the range and severity of the impairments that lead to social disadvantages. So, from an AT design and development perspective, the medical model can be more useful as the devices and services can be based on the functional limitations of their intended users. The main fallacy of this reasoning, as further shown shortly, is the incorrect assumption of the equivalence of impairment and disability. Furthermore, while such designs can address some of the practical issues that the seniors or people with disability face, research and development efforts with such convictions underpin a one-sided approach, placing the responsibilities to resolve access and inclusion issues especially on the individual [37].

Eyeglasses or contact lenses are examples that can be relevant to these discussions. In most case, these are not even considered as AT. While this is also related to social acceptability that will be discussed in Sect. 6.3, it has been used to exemplify some of the impacts that the disability views may have [27]. In many parts of the world, and definitely in developed countries, short-sightedness is not considered to be the cause of disability. People can easily and affordably get eye tests and corrective lenses to correct for that. So, such an impairment will not preclude an individual from fully participating in social life and pursuing their goals. The level of inclusion can be drastically different in some developing countries, where eye tests or corrective lenses are not readily available or are prohibitively unaffordable. An individual in those circumstances may not even learn how to read and write properly, stopping them from being involved with many other social activities. As a consequence of this impairment, the individual lives with difficulties in functioning and as such, based on ICF definitions, they are living with disability. In this case, the disability clearly stems from how advanced the society is. This example can also indicate how the social model can affect the views on disability and what needs to be done to overcome the physical, mental, economic, and institutionalized barriers aiming to provide opportunities for all people on an equal basis. Such a view has been argued to mean that "taken to its logical conclusion, there would be no disability within a fully developed society." [27].

2.3 Age and Disability Experiences

As pointed out in the previous section, there are divergent views on disability models. To some extent, the experts have also voiced different viewpoints on which model is of more value for the design of AT. Given the nature of this work, the discussions about these central topics are brief. However, a cross section of ideas with relevant references for the interested readers are included. In this section, the relation of those models and the requirements for improving aging and disability experiences are further explored. These experiences vary significantly from one person to another. The inaccessibility of environments that create disability and frustration for the seniors can be changed as the result of legislation, policy amendments, or technological advancements. The focus of this book is on the last one of these.

From the discussions so far, it is clear that a variety of approaches and models need to be in play. Irrespective of that, the CRPD has provided several principles to guide how the requirements can be addressed. These can be summarized as the following points [1]. They can be easily extended to include the aging experience and meeting the needs or desires of the elderly individuals.

- Services should be provided in the community, not in segregated sites.
- While many people with disability or elderly persons may need assistance to achieve a high quality of life and to participate in social activities on an equal basis with others, the individuals must be involved in the decisions about the services they receive and remain in full control over their lives.
- With the majority of services coming from family members or social networks as informal care, state funding of responsive formal care and services as a critical enabler of the full participation of the elderly or persons with disability needs to be mandated.
- Support and assistance are just the means to enabling the individual to live with self-sufficiency and social inclusion.

People living with disability or dementia and the elderly individuals face many challenges and have several concerns. Identifying them can help with ascertaining the requirements for AT. Some of these challenges appear to be common among many people. They include not surprisingly, mobility and transport concerns, wayfinding related issues, communicating with others, and access to information [38]. Narrowing down more specifically to the concerns for the elderly in some EU countries, similar ones, but perhaps with somehow different emphasis have been identified [39]. The major identified concerns were being capable of getting around, and seeing, hearing, and communicating with others. These concerns were followed by mental health and well-being and having the ability for self-care. In this survey, the elderly, have made particular mention of online banking and the use of the Internet among the difficulties they face. While the usage of the computer varied, it generally involved the use of the Internet, to search for new information, to stay involved with others, online banking, and playing games (http://www.i-stay-home.eu/).

Accessible communication systems and access to information and the Internet are increasingly important for full participation in the society. Studies have indicated that people with disability use ICT at considerably lower rates compared to the general population, which in some cases, included not even having access to services like the telephones, televisions, and the Internet [1]. People with hearing, speech, or vision impairment are significantly disadvantaged. However, studies have also shown that properly configured computer systems can be accessible even for deafblind [40]. The computer would then be a great asset for the daily living of these people. Individuals with a sensory impairment may use AT to access the Internet and the World Wide Web. While some sites or pages may remain or become inaccessible, with various initiatives already in place, the Web is becoming more accessible as time goes on [41].

Communication technology can be the preferred means of staying in touch with friends and family, easier than traveling to see them, for many elderly. Mobile phones, and increasingly smartphones, provide many capabilities for increased independence. Some studies have indicated that the elderly carry their mobile phones, mainly because they want their family members or carers to be able to contact them (https://www.ef-l.eu/assets/I-stay@home_complete2015.pdf). This study has found that smartphone users, however, employ the technology for games and financial management. This usage is obviously in line with that of the general population.

Another aspect which is also in line with the general population is the acceptance and employing of technology based on how others perceive the user. Some investigations have shown that while people with disability feel empowered when using AT, they want to avoid misperceptions by others [42]. For instance, some individuals prefer not to use white canes or hearing aids, to elude the stigma of being less capable. It is interesting to note that similar stereotypes are not associated with more contemporary digital devices. As it will be discussed in more detail in Sect. 6.3, this may relate to aesthetic of the AT and how dissimilar it looks from a functionally similar mainstream device. The low adoption of noticeable or distinct products may also explain why some elderly may avoid using unfamiliar technology and sophisticated devices, as they may feel troubled about being seen as unable to use the technology with ease, which could contribute to misperceptions that they were not capable.

Aesthetics requirements of AT devices can be simply overlooked by designers and professionals. Such inattentions are among the aspects where the dissimilarities or interactions between medical and social models of disability show themselves. Eyeglasses are widely used for correction of some vision impairment. Their success is partly attributed to them being generally viewed as fashion accessories, rather than AT [43]. Aesthetics, choice, widespread use, and similarity to sunglasses that are used by the general public have contributed to the acceptance of eyeglasses, by people with visual impairment. Utilization and usability by the public at large can also be related to the more general concept of design for all or universal design.

The idea behind the design for all concept, which incidentally is not that new, is to encourage the development of devices, products, and services that all people can use [44, 45]. This approach can improve affordability and avoid stigmas. Design for all ideas are not targeted towards the elderly or people with disability. The aim is to include as many of different people as possible, irrespective of their age, health status, gender, and other factors. This concept while well understood, is progressing at different paces in different industries. The increasing number of the elderly and their related large market and wishes to access buildings and environments, and to use public transport, smartphones, computers, and other products and services is a strong force behind this concept. The businesses can also benefit from using these concepts and improve their market shares. For instance, there are reports of a UK supplier whose £35,000 cost of making its online shop accessible for visually impaired people, resulted in £13 million a year trade through the site, mainly due to fully sighted customers finding it easier to use, compared to other sites [1]. Design for all principles have impressively influenced computer systems. Windows and Linux machines come with many accessibility options that facilitate their better use by the elderly or people with disability [41].

It is also important to consider the requirements of the services relevant to disability and the aging population. For example, while 80% of the aged care in high-income countries come from families, the effects of the aging population on the swelling of demand and the falling of supply of care, can be alarming [1]. As the majority of care for the elderly and people with disability or dementia is through informal care, meeting their requirements and needs can have significant burden on spouses, family members, or relatives. The burden of care makes responding to the needs of the caregivers and developing AAL and AT to support them mandatory. These people can also provide positive input for better design and development of AT and improve their adoption and uptake. The carers, as well as the people with disability and the elderly, have invaluable insights that can help with the understanding of the personal, social, environmental, and other parameters that can affect the development and uptake of the AT.

It also needs to be noted that AT for recreational and social activities are required and well-received by their intended users [19]. In particular, technologies that can help with identifying and participating in activities within the elderlies' social environments, can be inspirational and improve their social interactions and inclusion [46]. The technology setting requirements for the elderly may not be that different from those for the general population. The analysis of some research data has suggested that user-friendly and robust operations, enhanced visibility, ease of device mobility and portability, and extensibility top the list of the requirements by the elderly [47]. While the older adults may be particularly interested in health monitoring technologies, they have been found to be willing to use all AT, as long as they are easy to use [39].

Monitoring of health and other conditions can be critical for various reasons. For instance, the progress of dementia can be slowed if diagnosed in its early stages. Family members, or others living with an affected individual, are the first ones to notice the symptoms of dementia. For the elderly living by themselves, however, these signs can be easily missed [48]. Technology, particularly more advanced IoT-based smart environments, can be of significant value here. Obviously, these detections and diagnoses can be extended to many other symptoms or conditions, including falls, wandering, risky or dangerous situations, and others. Section 5.1 will present samples of many exciting works that aim to develop and implement the relevant technologies.

For successful adoption of AT, another requirement that research has identified, relates to the need for considering individual's unique characteristics and personality [12]. These features may also relate to the state of mind of the individual, how confident or anxious they are, and how they conceive others may react to their use of AT. In this sense, AT is adding to an already complex experience of the elderly and people living with disability [31]. There are also other challenges in adoption of AT, to be detailed in Sects. 5.2 and 6.3. For instance, while the main barrier for the uptake of AT is found to be affordability, lack of awareness of available AT and other options, and usability issues are not that far behind (https://www.ef-l.eu/assets/I-stay@home_complete2015.pdf).

The discussions so far, indicate that many devices and technologies can enhance the quality of life for many aged people and those living with disability or dementia. The main obstacles that need to be addressed relate to functionality, affordability, increasing awareness of what is already available, aesthetics, and usability. Some of these can be dealt with through involving all stakeholders in the aged care or disability experiences and moving towards a holistic approach where AT and AAL schemes and developments utilize design for all concepts while taking social parameters and individual characteristics into genuine consideration.

2.4 Assistive ICT and IoT

ICT is restructuring many societies, economies, and even how individuals live. With its huge market and diversity of uses, it is not easy to size the IT market. Some research has put the global IT market revenue from hardware, software, telecommunications, and related services to more than US$3.7 trillion in 2015 and US$3.8 trillion in 2016 (https://www.comptia.org/resources/it-industry-outlook-2016-final). The influences of the Internet, in particular, have been dramatic. Easier access to information, banking, shopping, consultations, and many other online services, communications and virtual visits with relatives and friends, and remote applications for education or health are just some of the well-developed benefits of the ICT and the Internet. There are also, somehow newer areas, where smart living and environments powered by the IoT, can offer advantages for many people.

Access to information online can be particularly of interest to elderly and some of the people living with disability. It may help with avoiding some of the physical, transport, and communication barriers of information access through other means. For instance, 95% of people with mental health conditions have indicated that the Internet is their source of diagnostic-specific information [49].

Most mainstream products and services, including the mobile phones and the Internet, are not compatible with assistive technology and devices like hearing aids or travel aids for the visually impaired people. However, this is rapidly changing. For instance, the industry and many governments are behind the Global Accessibility Reporting Initiative that can help people find suitable communication devices, like mobile phones or smart TVs, for various vision, hearing/speech, cognitive, and dexterity abilities in different regions of the world (http://www.gari.info/). Priority Assistive Products List (APL) is another global initiative that intends to improve access to assistive devices and products for everyone (http://www.who.int/phi/implementation/assistive_technology/global_survey-apl/en/). The APL is the first of four tools to be developed by the Global Cooperation on Assistive Technology (GATE) initiative (http://www.who.int/disabilities/technology/gate/en/). GATE is a WHO flagship program, which is being advanced in cooperation with other UN agencies and many other partners. APL and GATE aim to increase access to high-quality affordable AT for the seniors and individuals with a disability to enable them to have healthy, productive, and dignified lives. As pointed out in the previous section, there are strong business cases for the inclusion of all people and improving the usability of the mainstream devices and services by incorporating design for all concepts. There are also some solutions that specifically address a number of the requirements that were discussed in Sect. 2.3. The details of some of those solutions will be presented in Sects. 3.1, 3.4, and 5.1. A brief overview here starts those discussions.

Worldwide, it is estimated that there are 6.8 billion mobile phone subscriptions (http://data.worldbank.org/indicator/IT.CEL.SETS.P2). The high accessibility and widespread use of mobile phones or smartphones can offer many advantages through the development of apps and non-conventional services. For example, apps can make smartphones useful for screening of hearing loss, even though the measurement of the level of hearing impairment may not be accurate [50]. Moreover, smartphones and many other portable devices, with their significant computing power, can run many sophisticated applications very efficiently. These include various applications that are of value to the elderly or people living with disability. Many navigational systems, text-to-speech, and speech-to-text apps, with virtual audio displays or combinatory Braille and speech interfaces, can be conveniently integrated with most smartphones [51]. Some apps make the mobile devices capable of supporting Braille displays that can be connected to the smartphone using Bluetooth. Many apps that expand the AT-like capabilities of devices using both Android and iOS have already been developed and are available. Regarding accessibility, VoiceOver is among the substantive parts of the iOS devices (http://www.apple.com/accessibility/). It can be used by a visually impaired individual for reading the screen, navigation control in Safari web-browser, controlling the device, or taking photos.

For all of us, independently moving around and traveling is essential for education, employment, leisure, and participation. To travel around, blind and visually
impaired people may use AT for localization, obtaining orientation information,
avoiding obstacles, wayfinding, and other related purposes. While there are many
research and development works in these areas, such solutions in a complete and
widely usable form, do not exist yet [52]. Outcomes of various technologies like
wireless sensor networks, robotics, computer vision, Global Positioning System
(GPS), cellular and wireless system, and the IoT are being combined, aiming to
achieve the purposes of such AT. Some of the main functions of such AT and devices
can be summarized as navigation and Electronic Travel Aids (ETA).

The general public uses navigation systems to a great extent. The developments of
GPS are mainly the result of such a large market. The GPS and perhaps more so,
other navigation-related technology, like LIght Detection And Ranging (LIDAR), are
gaining even more importance due to the widespread growth of interest in autonomous cars (http://news.mit.edu/topic/autonomous-vehicles). Significant differences,
such as the required accuracy and audio and tactile interface requirements, do exist
between the devices that are suitable for use as an AT for visually impaired people
and those for other applications [53]. Nevertheless, the progress of these technologies and related devices can of value to all areas. Innovations, price reductions, and
many significant developments are related to the large market for these technologies
that can be of great benefit to all people, including for use as part of ETA for visually
impaired individuals and in other assistive devices. These developments and the use
of other technologies for such purposes are further discussed in Sects. 3.3 and 5.1.

Technology has changed the way we manage our daily lives. Again, partly due to
market forces, there are many devices and appliances available that can be used to
improve daily living for people with different ages and abilities (http://www.able-
data.com/). For instance, the talking microwave ovens that provide spoken information on various functions and status of the device can be of value to visually impaired
individuals   (https://www.visionaustralia.org/shop/product-list/product-detail/
cobalt-talking-microwave). There are also a variety of commercial services and
devices developed for network-based monitoring of movements, fitness, and well-
being of people in home care [54]. Talking ATMs are in use in most developed
countries and many developing countries. In many countries, including the US, the
provision of these ATMs is required by legislation and regulations (http://www.
lflegal.com/2010/09/doj-revised-regs/#atm).

Technology can also be valuable for promoting independent living, calmness,
living actively, facilitating participation, and providing reminders for people living
with dementia [55]. For instance, compared to traditional reminders, like a diary,
active reminders that function as an AT can be especially helpful for those with
memory loss or dementia [56]. Mobile technology and GPS can be used to locate
people, which can overcome the problems associated with wandering and getting
lost that are common among people living with dementia. A 3-year study, carried
out in 19 Norwegian municipalities with the participation of 208 individuals with
dementia, has shown the effectiveness of using GPS in maintaining their independence while safely continuing with outdoor activities [57]. This study has also
shown the benefits of this rather straightforward and inexpensive approach for their

family and caregivers. A systematic review of 12 studies has also demonstrated that the Internet interventions are particularly of significant benefit for the informal dementia caregivers, in terms of their competency, confidence, and managing depression [22].

The use of robots for aged care and as AT for assisting the elderly or people living with disability has also been suggested. According to International Federation of Robotics (IFR), typical applications include robotized wheelchairs, rehabilitation robotics, and the related smart prostheses and orthotics (http://www.ifr.org/news/ifr-press-release/professional-service-robots-continued-increase-412/). IFR sees growth in the area of rehabilitation robotics as a result of aging. However, they do not anticipate a substantial growth in the market for robots as AT for people with disability. A study in Japan has concluded that in their case, for older adults smooth operation of the robot was not achieved and therefore, it was not suitable for supporting the elderly who live by themselves [58].

References

1. World Health Organization. (2011). *World report on disability*. World Health Organization.
2. World Health Organization. (2016). Deafness and hearing loss. Available online, http://www.who.int/mediacentre/factsheets/fs300/en/ [Accessed 10-Oct-2016].
3. Hersh, M. A. (2008). Perception, the eye and assistive technology issues. In *Assistive Technology for Visually Impaired and Blind People* (pp. 51–101). Springer London.
4. Lancioni, G. E., & Singh, N. N. (2014). Assistive technologies for improving quality of life. In *Assistive Technologies for People with Diverse Abilities* (pp. 1–20). Springer New York.
5. Holton, B. (2016). A Day in the Life: Technology that Assists a Visually Impaired Person Throughout the Day. Available online, http://www.afb.org/info/living-with-vision-loss/using-technology/using-a-computer/123 [Accessed 4 March 2017].
6. Australian Bureau of Statistics (2015). Disability, ageing and carers, Australia: Summary of findings, Available online: http://www.abs.gov.au/ausstats/abs@.nsf/mf/4430.0 [Accessed 4 March 2017].
7. United Nations Department of Economic and Social Affairs, Population Division (2015). World Population Ageing 2015 (ST/ESA/SER.A/390). Available online, http://www.un.org/en/development/desa/population/publications/pdf/ageing/WPA2015_Report.pdf [Accessed 4 March 2017].
8. United Nations Department of Economic and Social Affairs, Population Division (2015). World Population Prospects: The 2015 Revision. Custom data acquired via website, https://esa.un.org/unpd/wpp/ [Accessed: 10-Oct-2016].
9. United Nations Department of Economic and Social Affairs, Population Division (2015). World Population Prospects: The 2015 Revision, Key Findings and Advance Tables. Working Paper No. ESA/P/WP.241.Available online, https://esa.un.org/unpd/wpp/publications/files/key_findings_wpp_2015.pdf [Accessed 4 March 2017].
10. Caprani, N., O'Connor, N. E., & Gurrin, C. (2012). Touch screens for the older user. In *Assistive technologies*. InTech.
11. Taylor, H. R., & Keeffe, J. E. (2001). World blindness: a 21st century perspective. *British Journal of Ophthalmology, 85*(3), 261–266.
12. Federici, S., Tiberio, L., & Scherer, M. J. (2014). Ambient Assistive Technology for People with Dementia: An Answer to the Epidemiologic Transition. *New Research on Assistive Technologies: Uses and Limitations. New York, NY: Nova Publishers*, 1–30.

13. World Health Organization. (2015). World report on ageing and health. Geneva: World Health Organization. Available online, http://www.who.int/ageing/publications/world-report-2015/en/ [Accessed 10-Oct-2016].

14. Prince, M. J. (2015). *World Alzheimer Report 2015: the global impact of dementia: an analysis of prevalence, incidence, cost and trends.*

15. Singh, N. N., Lancioni, G. E., Sigafoos, J., O'Reilly, M. F., & Winton, A. S. (2014). Assistive technology for people with Alzheimer's disease. In *Assistive technologies for people with diverse abilities* (pp. 219–250). Springer New York.

16. Takahashi, Y., Kawai, T., & Komeda, T. (2014). Development of a Daily Life Support System for Elderly Persons with Dementia in the Care Facility. *Studies in health technology and informatics, 217*, 1036–1039.

17. United Nations Division of Social Policy and Development Disability. (Online). Monitoring and Evaluation of Disability-Inclusive Development. Available online: https://www.un.org/development/desa/disabilities/resources/monitoring-and-evaluation-of-inclusivedevelopment-data-and-statistics.html [Accessed 4 March 2017].

18. United Nations Development Program. (2017). Disability Rights. Available online, http://www.undp.org/content/undp/en/home/ourwork/povertyreduction/focus_areas/focus_inclusive_development/disabilityrights.html [Accessed 4 March 2017].

19. Hersh, M. A., & Johnson, M. A. (2008). Assistive technology for education, employment and recreation. *Assistive Technology for Visually Impaired and Blind People*, 659–707. Springer London.

20. Brown, E. (2016). Disability awareness: The fight for accessibility. *Nature, 532*(7597), 137–139.

21. Björk, E. (2012). *Universal Design Or Modular-Based Design Solutions-A Society Concern.* INTECH Open Access Publisher.

22. Boots, L. M. M., Vugt, M. E., Knippenberg, R. J. M., Kempen, G. I. J. M., & Verhey, F. R. J. (2014). A systematic review of Internet-based supportive interventions for caregivers of patients with dementia. *International journal of geriatric psychiatry, 29*(4), 331–344.

23. Panou, M., Cabrera, M. F., Bekiaris, E., & Touliou, K. (2014). ICT services for prolonging independent living of the elderly with cognitive impairments-IN LIFE concept. *Studies in health technology and informatics, 217*, 659–663.

24. Alzheimer's Asoociation. (2015). 2015 Alzheimer's disease facts and figures. *Alzheimer's & dementia: the journal of the Alzheimer's Association, 11*(3), 332.

25. Friedrich, M. J. (2015). New figures on global dementia cases. Jama, 314(15), 1553–1553.

26. Li, R., Lu, B., & McDonald-Maier, K. D. (2015). Cognitive assisted living ambient system: a survey. *Digital Communications and Networks, 1*(4), 229–252.

27. Llewellyn, A., & Hogan, K. (2000). The use and abuse of models of disability. *Disability & Society, 15*(1), 157–165.

28. World Health Organization. (2001). *International Classification of Functioning, Disability and Health: ICF.* World Health Organization.

29. Shakespeare, T., & Watson, N. (2001). The social model of disability: an outdated ideology?. In *Exploring theories and expanding methodologies: Where we are and where we need to go* (pp. 9–28). Emerald Group Publishing Limited.

30. Corker, M., & Shakespeare, T. (Eds.). (2002). *Disability/postmodernity: Embodying disability theory.* Bloomsbury Publishing

31. Frauenberger, C. (2015, October). Disability and technology: A critical realist perspective. In *Proceedings of the 17th International ACM SIGACCESS Conference on Computers & Accessibility* (pp. 89–96).

32. United Nations General Assembly (2006). Convention on the Rights of Persons with Disabilities. Geneva, *GA Res, 61*, 106. Available online, https://www.un.org/development/desa/disabilities/convention-on-the-rights-of-persons-with-disabilities.html [Accessed 11 April 2017].

33. Shakespeare, T. (2013). *Disability rights and wrongs revisited.* Routledge.

34. Söder, M. (2009). Tensions, perspectives and themes in disability studies. *Scandinavian journal of disability research, 11*(2), 67–81.
35. Nihei, M., & Fujie, M. G. (2012). *Proposal for a New Development Methodology for Assistive Technology Based on a Psychological Model of Elderly People*. INTECH Open Access Publisher.
36. Gilliard, J., Means, R., Beattie, A., & Daker-White, G. (2005). Dementia care in England and the social model of disability lessons and issues. *Dementia, 4*(4), 571–586.
37. Mankoff, J., Hayes, G. R., & Kasnitz, D. (2010, October). Disability studies as a source of critical inquiry for the field of assistive technology. In *Proceedings of the 12th international ACM SIGACCESS conference on Computers and accessibility* (pp. 3–10).
38. Coetzee, L., & Olivrin, G. (2012). *Inclusion through the Internet of Things*. INTECH Open Access Publisher.
39. Williamson, T. (2016). Review of the I-Stay@ home (ICT Solutions for an Ageing Society) study Wiki. *Journal of Assistive Technologies*.
40. Hersh, M.A., and Johnson, M.A., 2005, Information technology, accessibility and deafblind people, proceedings of Association for Advancement of Assistive Technology in Europe Annual Conference, Lille, France.
41. Hersh, M. A., & Johnson, M. A. (2008). Accessible information: an overview. *Assistive technology for visually impaired and blind people*, 385–448. Springer London.
42. Shinohara, K., & Wobbrock, J. O. (2011, May). In the shadow of misperception: assistive technology use and social interactions. In *Proceedings of the SIGCHI Conference on Human Factors in Computing Systems* (pp. 705–714).
43. Pullin, G. (2009). *Design meets disability*. MIT press.
44. Mace, R.L. (1985). *Universal Design, Barrier-Free Environments for Everyone*, Designers West, Los Angeles.
45. Preiser, W. F., & Ostroff, E. (2001). Universal design handbook. McGraw Hill Professional, 2001.
46. Moritz, E. F., Biel, S., Burkhard, M., Erdt, S., Payá, J. G., Ganzarain, J., & Cabello, U. V. (2014). Functions: How we understood and realized functions of real importance to users. In *Assistive Technologies for the Interaction of the Elderly* (pp. 49–68). Springer International Publishing.
47. Man, Y. P., Cremers, G., Spreeuwenberg, M., & de Witte, L. (2015). Platform for frail elderly people supporting information and communication. *Studies in health technology and informatics, 217*, 311.
48. Abe, Y., Toya, M., & Inoue, M. (2013, October). Early detection system considering types of dementia by behavior sensing. In *Consumer Electronics (GCCE), 2013 IEEE 2nd Global Conference on* (pp. 348–349).
49. Cook, J. A., Fitzgibbon, G., Batteiger, D., Grey, D. D., Caras, S., Dansky, H., & Priester, F. (2005). Information Technology attitudes and behaviors among individuals with psychiatric disabilities who use the internet: Results of a Web-based survey. *Disability Studies Quarterly, 25*(2).
50. Abu-Ghanem, S., Handzel, O., Ness, L., Ben-Artzi-Blima, M., Fait-Ghelbendorf, K., & Himmelfarb, M. (2016). Smartphone-based audiometric test for screening hearing loss in the elderly. *European Archives of Oto-Rhino-Laryngology, 273*(2), 333–339.
51. Csapó, Á., Wersényi, G., Nagy, H., & Stockman, T. (2015). A survey of assistive technologies and applications for blind users on mobile platforms: a review and foundation for research. *Journal on Multimodal User Interfaces, 9*(4), 275–286.
52. Xiao, J., Joseph, S. L., Zhang, X., Li, B., Li, X., & Zhang, J. (2015). An assistive navigation framework for the visually impaired. *IEEE Transactions on Human-Machine Systems, 45*(5), 635–640.
53. Hersh, M. A., & Johnson, M. A. (2008). Mobility: an overview. *Assistive technology for visually impaired and blind people*, 167–208. Springer London.

54. Clark, J. S., & Turner, K. J. (2016). Evaluating automated goals for home care support. *Journal of Assistive Technologies*, *10*(2), 79–91.

55. Riikonen, M., Paavilainen, E., & Salo, H. (2013). Factors supporting the use of technology in daily life of home-living people with dementia. *Technology and Disability*, *25*(4), 233–243.

56. Baric, V. B., Tegelström, V., Ekblad, E., & Hemmingsson, H. (2015). Usability of RemindMe– An Interactive Web-Based Mobile Reminder Calendar: A Professional's Perspective. *Studies in health technology and informatics*, *217*, 247.

57. Øderud, T., Landmark, B., Eriksen, S., Fossberg, A. B., Aketun, S., Omland, M., & Ausen, D. (2015). Persons with Dementia and Their Caregivers Using GPS. *Studies in health technology and informatics*, *217*, 212.

58. Inoue, K., Sasaki, C., & Nakamura, M. (2014). Communication Robots for Elderly People and Their Families to Support Their Daily Lives-Case Study of Two Families Living with the Communicaton Robot. *Studies in health technology and informatics*, *217*, 980–983.

Chapter 3
Digital Senses and Cognitive Assistance

Institute of Electrical and Electronic Engineers (IEEE) and one of its Future Directions team have initiated work on "digital senses" based on technologies that relate to various human sensory systems [1]. One of the major objectives of this initiative is to promote alliances in the three focus areas of virtual reality, augmented reality, and human augmentation. Taking similar views, this part aims to show how combining the understandings about human senses with the advances in sensor technologies and the permeating nature of the IoT, can be of great value for aged care, cognitive assistance, and provision of disability services. As discussed in the previous section, the primary aim of these assistive devices and technologies is to enhance the functioning, independent participation, and overall quality of life for individuals. Some samples of such technologies, along with examples of the devices and systems that can be used in these contexts, are also discussed in this chapter. The discussions start with the traditional AT, then cover cognition assistance, and improvements of these two AT types through sensing to reach context-aware individualized ambient assisted living.

3.1 Conventional Assistive Devices and Technologies

The most commonly cited definition of AT is based on the Technical Assistance to the States Act. It considers AT as "any item, piece of equipment, or product system, whether acquired commercially off the shelf, modified, or customized, that is used to increase, maintain or improve functional capabilities of individuals with disabilities." The express purpose of AT is to facilitate access to environment and services [2]. Assistive devices include wheelchairs, prostheses, specially designed pens, hearings aids, visual aids, other hardware, and software that improve ICT access or communication capacities [3].

© Springer International Publishing AG 2017
S. Shahrestani, *Internet of Things and Smart Environments*,
DOI 10.1007/978-3-319-60164-9_3

The large range of AT aims to improve life quality for individuals with some sensory, cognitive, or mobility impairment. As mentioned in Sect. 2.1, with estimated 285 million visually impaired people and 360 million hard of hearing individuals, the most commonly encountered sensory impairments, are those related to hearing and sight. Age-related hearing loss or presbycusis, most often affecting both ears, is also very common. However, for various reasons, some of which were outlined in Sect. 2.4, assistive devices do not appear to be employed as much as they should. Only 5–15% of people living in low to mid-income countries who need AT have access to them (http://www.who.int/disabilities/technology/en/). Even in the US, with more that 35 million people living with hearing impairments, only 30% of adults with hearing loss have used hearing aids [4]. While affordability can be a factor, other reasons related to poor benefits of use can be in play resulting in this low uptake.

The global market for hearing aids is expected to grow from US$6.2 billion in 2015 to more than US$8.3 billion by 2020 (http://www.marketsandmarkets.com/PressReleases/hearing-aids.asp). A hearing aid contains a microphone for converting sound into electrical signals, a user adjustable volume and amplification control, usually a telecoil, a battery compartment, a receiver or the loudspeaker, and in most cases an integrated circuit where the signal processing is implemented [5]. Hearing aids, with Digital Signal Processing (DSP) capabilities, have been available since 1996. They have many advantages over the analog aids and a few limitations. DSP has been implemented in 93% of the hearing aids sold in the US in 2005 [6]. A cochlear implant is a different electronic device for individuals with severe to profound hearing loss. It converts the sound into electrical impulses that directly stimulate the hearing nerve [7]. The past couple of decades has witnessed rapid and exciting progress in these areas. Affordability and usability still result in rather small adoptions by the intended users. The future of these devices is tied to research and developments in wireless technology, DSP and related chip miniaturization, cognitive science, and artificial intelligence [8]. As with most other devices and applications, management of individual needs and being capable of meeting them will drive their success and employment.

Mobility and capability of independently traveling around are among the most fundamental needs for all of us, including for blind or visually impaired individuals. While the visually impaired people have used some form of stick for probing for centuries, the long cane has come into widespread and systemic use after the second World War and in the early 1950s [9]. The use of the guide dogs goes back to 1920s. While a guide dog is not an AT, due to its effectiveness for purposes similar to those for AT, its analysis and study can provide useful insight for technology design and implementations [10]. The long cane, a simple standard mechanical mobility device, appears to be the most effective AT for independent travel by blind people. However, some may prefer to avoid the labeling that goes with it. It can also be quite tiring to use. Recent electronic versions of the cane, equipped with sensors and other pieces of technology, will be discussed in the next sections. Their adoption, compared to the low-tech long cane, remains a questionable matter. More new AT that can be of value for orientation and spatial sensing, to be discussed subsequently, include ultrasonic systems, variations of GPS, talking signs, systems based on cellular and wireless networking technologies, and IoT-based AT.

Reading and writing text are also the abilities that are essential for participation. Visually impaired individuals can use either listening or haptic abilities for reading. They can listen to talking books, use available text-to-speech technology, or Braille for reading. The converse can be used for writing. In relation to the elderly, it may be important to note that some studies have indicated that many older people have difficulty in learning and using Braille, as their tactual perception is not as good as that of a young person [11].

For cases of cognitive conditions, the use of AT has not been as consistent as it is for sensory impairments. In particular, the use of conventional AT by people living with dementia has been very limited [12]. This limited use has been attributed to the view that the affected individuals cannot use AT at levels that can make a difference to their lives [13]. However, there are clear reports of the use of technologies that have improved the functioning or have meaningfully enriched the living environments of people with dementia [14]. For instance, in Sect. 2.4, the use of GPS for alleviating some of the issues associated with wandering and disorientation has been mentioned. Another example is that of using the benefits of the relatively intact autobiographical memory of the individual to enrich the reality for them [15]. This approach does not require much learning. In essence, it is based on collecting the autobiographical memory data that are then displayed on memory wallets, smartphones, or computers to engage the individuals in meaningful conversations and interactions.

Wandering is a characteristic of some individuals with AD dementia that can be quite worrying and exhausting for their carers [16]. AT to help caregivers with this issue is available. For example, CareWatch can be used to monitor if the unattended individual has exited through a door or if they have come off their bed. Some studies have indicated their value, as well as the benefits of some other relevant AT, in improved "peace of mind" for the caregivers [17]. Some watches are available that can be used by all, including the elderly, to alert an aid-worker or others in cases of an emergency or needs for help (http://www.thecarewatch.com/).

An important dimension of dementia is the provision of support for caregivers. It is well established that informal care to assist older adults with dementia can cause physical, emotional, social, and financial issues for the caregiver [18]. As such, one of the major points in suitable dementia and aged care, is related to proper design and implementation of AT to support the carers. There have been several works reported in this area. For example, a 4-year long study of AT designed for dementia has shown that most carers valued the positive results from using the AT devices, with more than one-third reporting that the AT use mitigated the emotional burden of worry [19, 20]. The majority of the AT for dementia, actually offer some support and assurances for the carers, rather than provide direct assistance for the person living with dementia [21]. For example, the use of mobile phones, wander alarm, or GPS-based systems are mostly as a courtesy or reassurance for the caregiver.

3.2 Cognition Models and Assistance

Human cognition and intelligence are not well understood. Most researchers consider human cognition capabilities falling into two major types, namely analytic and synthetic cognition. Analytic cognition is based on abstraction and separating an object from its context, attempting to categorize it using its characteristics, aiming to explain and predict the behavior of the object using generalized rules about the categories [22]. This type of cognition involves the use of mathematical skills, formal logic, algorithmic processing, and symbols. These can be suitable processes for devices or systems with some computational intelligence. Synthetic cognitive capabilities, on the other hand, relate to creativity and can incorporate intuitions, emotions, and past experiences. This type of cognition does not attempt to categorize objects and does not aim to provide a model that can explain or predict the object behavior [23].

As with cognition, there is no widely accepted definition of intelligence and Artificial Intelligence (AI). The latter refers to machines and human-made systems that can accomplish tasks that normally require skills and abilities associated with human intelligence. These abilities include functions related to learning, visual and speech recognition, decision-making, and planning. With recent huge leaps made possible by deep learning, big data, and analytics, World Bank sees AI as one of the six major technologies to watch for their growth and influences on our lives [24]. For quite some time now, AI has been able to assist physicians with diagnostics [25], while recent advances can make cars capable of driving themselves on busy highways (https://waymo.com/). Siri from Apple, Android SpeechRecognizer, Amazon Alexa, and Microsoft Cortana are among several other AI-based systems providing near perfect and practical voice recognition that is easy to use for most people. These voice recognition and natural language processing capabilities plus cloud-based reasoning and complex AI systems, which are accessible through smartphones and similar readily-available devices, can really make the difference for personalized care and AT. With the huge market for voice-controlled technologies, beneficial for all smartphone users, the already extensive range of available products and services can only expand with time. Such expansions can be of great practical benefit for assistive devices and technologies. Particularly, the mainstream nature of most of these products and services can be of value in their uptake by many potential AT users.

Traditional AI approaches can efficiently utilize rule-based systems and algorithms. That is, they can be efficiently based on analytic cognition [26]. In human beings, cognition capabilities of analytic type, appear to be the ones that deteriorate early with aging and dementia. On the other hand, for most humans the synthetic cognition seems to come more innately and effortlessly, helping people with learning from many examples rather than developing models, staying with them for longer. This cognition is closer to the newer concepts of deep machine learning [27]. Realizing these distinctions and implementing the recent AI advances, the scope of the cognition assistance for aged care and people living with disability or dementia can be tailored to meet each individual's needs.

Most of the works that use AI to produce cognitive orthoses for people with deteriorating or different abilities are still in the development stages [21]. Some samples of the relevant studies and their outcomes are presented here. An AI-based complex AT, aiming to assist individuals with dementia to do specific daily activities has shown some promising results [15]. The results of using the AT indicate that compared to the baseline, a typical activity was completed by 30% of the individuals with 11% more right steps, while the requests for support from the caregivers went down by 60%. Around 60% of the participants are reported to have achieved nearly full independence. In general, building a profile of typical living habits of the residents of a home and recognizing deviations from those habits can help with detecting a decline in health or cognitive abilities. Noting that around 30% of the elderly with dementia live alone, these detections can be principally significant [28]. AI can subsequently help with identifying the situations that interventions may be needed. These include approaches that are based on machine learning using the data collected by environmental and wearable sensors [29]. Rule-based, case-based, ontology-based, game-theoretic, and intelligent agent-based approaches have offered some solution for incorporation in AAL. For instance, to identify situations that require attention and to provide meaningful interventions when necessary, an agent-based system that monitors the human interactions with their environment has been proposed [30]. In this architecture, temporal specifications of the interactions and the events that are monitored, along with their deviations from previously established baselines recognized through cognitive analysis, are used to generate the signals for the required interventions.

More specifically for dementia, some studies have shown the benefits of engaging those diagnosed with AD in virtual reality programs [31]. This engagement is believed to improve the chances of the individuals in maintaining their daily life autonomy. These works focus on training individuals to engage in some tasks independently, by using their intact abilities and cognitive skills to replace the capabilities that are impaired. To overcome spatial disorientation, getting confused or lost, training in wayfinding in virtual reality can also be of value. The aim of this training is to reinforce the use of the autobiographical memory of the significant landmarks and to practice paths between them, rather than trying to remember new markers.

The positive effects of using the relatively intact autobiographical memory of the individual to enrich the reality can be expanded by noting that memories, photos, past experiences, or knowledge are not confined to the individual. Such benefits can be related to the fact that our cognition is part of our composite sociocultural environment and is constantly affected by it [32]. The concept of distributed cognition can explain, at least partly, why elderly or people with dementia prefer to stay in their homes and with familiar people. In these situations, the cognitive paths are distributed across these familiar surroundings and persons. Distributed cognition concepts have been used in AT developments. An example of using distributed cognition is a computer-based system that uses publicly available items to stimulate the autobiographic memory of the individual to improve their interactions with family and carers [33]. Their effectiveness in improving communications, participation, and interactions have been shown in its trials. The success of this system can be partially attributed to having all of its aspects including its contents, touch-screen, and other interfaces being developed with the involvement of the affected individuals, their relatives, and carers [34].

AAL can be of value to the elderly with various degrees of cognitive impairment, as well as to those with some other impairment and their caregivers. The development of AAL for aged care in home environments, in more general terms, is also significant. For instance, as part of the leading initiative Digital Agenda for Europe, the Active and Assisted Living Joint Programme has put AAL technologies at the forefront of Ageing Well in the Information Society action plan (https://ec.europa.eu/digital-single-market/en/news/new-official-website-ambient-assisted-living-joint-programme-aal-jp). IN LIFE is one of the European Commission co-funded projects (http://www.inlife-project.eu/). The project utilizes the well-developed notions of several pilots used to evaluate the technology readiness of connected services for the elderly. The main functionalities to be achieved by this project are, modestly monitoring users' related activities and preferences, supporting users in a range of indoor and outdoor activities, and providing assistance to caregivers [28]. The project goals also include providing easy access while maintaining user privacy and information security, and enabling easy integration of products and service from other providers.

3.3 Improving Assistive Devices with Sensing, Communication, and Actuation Capabilities

The previous sections outlined some of the conventional AT and the exciting technological advances that have or can make them more suitable and efficient for their purposes. In this section, some of the relevant ideas from other fields including robotics, computer vision, voice-activated technologies, voice to text, text to voice, and face recognition that can cross-feed into AT areas are discussed. Such deliberations can be prohibitively large for coverage in a book of this nature. With that in mind, the AT for visually impaired that encompasses various requirements, are focused on to exemplify the ideas, while other assistive devices and services are also considered. In line with the theme of this work, assistive IoT and smart environment benefits, this section concentrates on the potential enhancements of the AT with the addition of sensing and actuating capabilities.

For the improved quality of life and independent living, visually impaired individuals require AT to enable them to travel and move around. Indoor and outdoor movement and mobility have various challenges that are yet to be fully met by the available AT. These systems must sense and have access to various information, process them, and provide the results to the user through suitable interfaces. The technology needs to deal with localization and positioning, orientation, efficient wayfinding capabilities, dynamic obstacle avoidance, hazards warnings, notifications of events or people of interest, and several other requirements. Smart environments, sensor technologies, wearable devices, smartphones, and the IoT can play significant roles in meeting some of these necessities. The technology that can achieve all of the requirements does not appear to have been developed yet; but what is available can meet most of these needs. What is lacking is a holistic approach that can bring together various stakeholders and technologies to achieve suitable outcomes.

For visually impaired people, the substitution of visual information by tactile stimuli has been shown to be a viable solution for tasks like recognizing basic shapes, reading, and localization [35]. However, this substitution may require substantially more complex AT for use in navigation or functions of similar nature. Russell Path Sounder, invented in 1966, is one of the first sensor-based devices that can provide tactile and auditory warning of obstacles using ultrasonic beams [36]. This invention was followed by many modified or improved devices with similar functionalities. Mowast, for instance, enhanced the range of PathSounder from 2 m to around 4 m in 1972. The Laser Cane then made some improvements by using laser beams and receivers to perceive the distance from down curbs and obstacles, within a selectable range of 3–4 m forward. Another AT with similar purpose, is Kay Sonic Torch, with developments dating back to 1964. It is considered to be an environmental sensor, providing auditory stimuli that are very rich in information. Another practical and relatively simple AT that works with ultrasonic beams rebounding from nearby objects is the MiniGuide [37]. This advanced handheld sensor, shown in Fig. 3.1, can scan, detect objects, and provide auditory and vibrating information to the walking user. It is a small handheld programmable device suitable for various environments that is typically used along with the long cane or a guide dog.

An excellent overview of the old and the newer AT for the visually impaired and the blind, along with discussions on their usability can be found in [10]. While the developments of electronic mobility aids for the blind date back to 1897, the realization of the suitability of ultrasound and infrared radiation for remote sensing during and after WWII facilitated their serious progress. Since then, various devices and technologies have been developed. They are based on one or a combination of the

Fig. 3.1 MiniGuide, the image used with permission from GDP Research

following concepts and technologies: ultrasonic, sound signs, infrared lights, GPS, laser, cellular telephony systems, radio-frequency, wireless communications, and computer networks.

Many of the powerful solutions for navigation by visually impaired or blind people that have been proposed, remain incomplete or not widely adopted. Their main distinguishing factor is their goal for indoor versus outdoor navigation. A good summary of many of those solutions along with another developed proposal is given in [38]. This proposed solution is based on a platform that processes the information collected from the user and the environment to generate navigation messages for delivery to the individual as they move in an indoor area. The system uses a long cane augmented with infrared lights, two infrared cameras, a computing unit, and a smartphone that passes voice-based navigation messages to the user. The environmental information is collected in an XML file that the computing unit utilizes for its processes. The primary limitations of this approach are the shortcomings for encounters with new environments and changes, such as the movement of a piece of furniture, which will require the regeneration of the maps.

Radio-Frequency Identification (RFID) is a long-standing technique that uses radio waves to track and identify people, animal, objects, and shipments. RFID-based systems have enjoyed more acceptance than many other solutions to problems involving positioning and tracking. They usually constitute of some computing unit, an RFID reader, and tags that are positioned strategically. The system is directing the user to move through predefined paths, which may be achieved in different manners. For instance, identifying the position of the user with a tag [39] or guiding the user with passive RFID and smartphones have been proposed [40, 41], among many others. The RFID-based solutions have a major limitation that restricts the user to movements in areas with predefined routes. Furthermore, the short range of RFID can limit its employment in many situations of practical significance. Some works that combine RFID, GPS, and robots to guide the elderly or visually impaired individuals have claimed to overcome this limitation, at least to some extent [42–44].

There are also many advances in robotics, computer vision, cellular telephony, face recognition, wearable computing, and autonomous vehicles that can be of value to many aspects of advancing the AT and implementing the digital senses. In some cases, though, the size, used interfaces, power requirements, costs, and other similar concerns may limit their use in AT devices. For instance, mounting sensors, scanners, or readers is possible on cars or some wheelchairs but may restrict the user mobility if deployed on a long cane [38]. The rapid advances in technology can potentially overcome many of these issues.

The sensing capabilities of light detection and ranging, or LIDAR, and some of the applications that take advantage of these capabilities were discussed in Sect. 2.4. This sensor works conceptually similar to radar, but it is capable of much higher resolutions in detecting objects. The higher precision is related to the use of lights, with wavelengths in the few hundred nanometer range, compared to using ultrasound with a wavelength of 1.9 cm or so. The hundreds of thousands of times smaller wavelengths of LIDAR can measure the distance to each point of an object in front of it with tremendous precision. This information can then be used to create a three-dimensional model of the world around the sensor [45]. While the creation of such a model can be of tremendous value in many AT implementations, particularly those intended for visually impaired people, several challenges need to be addressed first. One of them relates to interfaces and devising of proper ways and links to pass the information in a useful manner to the human user. The other ones relate to size and cost of LIDAR. Most LIDARS currently have rather large moving mechanical parts and cost from US$1,000 to US$70,000 [46].

The cost and size of the commercially available LIDARs make them impractical for use in most AT applications. The good news is that LIDARs are undergoing dramatic transformations. For example, MIT researchers have announced promising results with the radical improvements being undertaken primarily for obstacle avoidance, targeting driverless car market [46]. The works on solid-state LIDAR chips can bring their cost down to around $10 each. These will be on chip devices that are robust with no moving parts, small, and light. Currently, MIT LIDAR systems can detect objects at ranges of up to 2 m, with centimeter longitudinal and expected 3cm lateral resolution. The aim is to achieve a 10-m range, within a year [47]. While that range may need to be expanded drastically for use in autonomous vehicles, it sounds to be quite suitable for many assistive devices. Such solid state LIDARs can provide many practical solutions to the issues mentioned earlier in the provisions of a form of digital sense for sight.

The use of Radio-Frequency (RF) signals in the design of AT and ETA, particularly for indoor navigation, is also a long-standing approach. An excellent ETA survey that covers such approaches is reported in [48]. Many other works also have some elements of using RF in ETA, including the use of RFID [43], visible light communications [49], Ultra-Wideband (UWB) technology [50], and wireless sensors [51]. It is interesting to note that many of these works and similar developments also utilize the smartphone and its sensors or interfaces for one purpose or another.

While, as they stand, all of these RF-based systems have serious drawbacks, some of them offer partly acceptable solutions that may be utilized for complementing other approaches. The RFID-based solutions, or similarly those based on NFC, for instance, have short ranges requiring many tags and readers. Other systems may require line of sight, which is not feasible in many real-life situations. Those relying on signal strengths and triangulation for positioning can be inaccurate and actually incorrect due to multipath propagation phenomenon. To counteract that shortcoming, it has been noted that the narrow RF pulses of the UWB technology allow for distinguishing between the direct and the reflected signals, making them suitable for achieving relatively precise positioning in indoor environments. The cost of the sensors and other dedicated resources, along with the challenging timing and synchronization requirements, are the major limiting factors for utilization of UWB-based solutions. The use of UWB-based navigation in conjunction with other sensors in the IoT-based environments is further explored in Sect. 5.1.

Adding sensing capabilities to an environment can also assist elders who may be at risk of dementia. As discussed in Sect. 2.3, timely detection of dementia symptoms can help with early diagnosis of the disease and decelerating its progress.

This detection can be of particular benefit to people at risk who live by themselves. The sensors that are installed in the home of the older adults can detect their behavior and identify those exhibiting any suspicious symptoms to alert them or their caregivers. For instance, failing to take a shower or to turn off a faucet can be related to memory loss. Changed patterns of sleep may indicate insomnia and sleep disorders, which in conjunction with other factors, if present, can raise the suspicion of early stages of some dementia [52]. Obviously, the analysis must be based on more complex combinations of behavior and symptoms that may be carried out remotely in the cloud using AI techniques.

An array of works has also been conducted to deal with issues other than just the mobility of visually impaired individuals. For instance, a smart doorbell that can recognize previously encountered individuals who are at the door can be mentioned [53]. It can then notify the elderly, or a person with visual, hearing, or mental impairment through the network or on their mobile device. This system can help a person with visual impairment by recognizing the visitor. It can assist a hearing impaired individual by detecting that a visitor has rung the bell, identifying them, and sending a notification to the individual. For elderly, it may avoid the discomfort of moving to the intercom unnecessarily.

Smartphones and the use of apps can also play significant roles in the ETA and, more generally, AT developments. Some apps and usages of smartphones in voice-controlled technologies were mentioned in Sect. 3.2. There are extremely fast and positive developments in this area. For example, vision-related detection features, including face and object recognition, analysis of emotional facial attributes, and text extraction and character recognition are conveniently possible through machine learning on Google CloudPlatform (https://cloud.google.com/vision/). Through Cloud Vision API, Google allows developers to add these features to their apps. Suitably developed apps that integrate such features can be of obvious benefits to the elderly in general, as well as to people with dementia or visual impairments. These apps can help the blind people with identifying close landmarks and events of interest, friends nearby, or reading labels and the like. A person with dementia may be able to use them for the identification of some of the people that they know, but do not clearly remember. In these contexts, reliable and fast face recognition apps can be clearly valuable. Agile and accurate enough apps can assist a blind person, for example, to recognize a friend in the street or a shopping center. There are many apps with these characteristics. One of them that perhaps because of its mischievous uses, as well as its quick search capabilities has been well publicized is FindFace [54]. The app, released in Russia, allows a user to snap a photo of any person which is then used to search through one billion photos in 1 second to identify that person with 70% reliability. Many other, mostly web-based, solutions, including Google Images (https://images.google.com/) can also efficiently address this so-called person search or reverse image search problem (https://facedetection.com/online-reverse-image-search/).

Many apps that assist or facilitate navigation already exist too. Some of them use Wi-Fi or cellular signals, the inbuilt and fitted sensors of the phone, or its processing power to provide many services that can be of benefit to all, the aged, or those living with disability or dementia. For example, many Android apps for straight walking, rough positioning, getting compass capabilities, GPS navigations, and more generally for improved independence and life quality for those living with disability and dementia have already been developed (https://play.google.com/). Apple uses the accessible design paradigm in its products and promotes it for app and accessory developers. Apple devices and iOS have many in-built accessibility interfaces and features (https://developer.apple.com/accessibility/ios/). The VoiceOver is a screen reader that can facilitate the interaction of the user with objects in various apps, even when they cannot see the interface itself. It can be used to configure captions and audio descriptions to suit the user's needs. The AVSpeechSynthesizer class can read the selected text in an app in over 30 languages. Through Guided Access, iOS can help people with some cognition, and sensory impairment stay focused on a task. These built-in features are complemented with an array of native and developed apps with many advanced features (https://itunes.apple.com/).

This section has mentioned some of the advances in several diverse areas that can be beneficial in AT developments. The diversity of these areas is yet another aspect that mandates the need for taking a holistic view in developing assistive services for aging, disability, and dementia. Smartphone with their apps and various gadgets, advances in communication technologies, AI and deep learning, robotics, sensor technology, and positioning systems are widely used by all people. Their integration with the customary and more traditional AT, may need to be considered as part of one solution to improve the adoption of the assistive devices and services by their intended users. These technologies, services, and devices can be part of the IoT and smart environments that will be discussed in more details in Sect. 5.1.

3.4 Ambient Intelligence and Assisted Living

Ambient Assisted Living (AAL) encompasses many diverse technologies with practical, industrial, and research significance. Smart environments and ambient intelligence, Wireless Sensor Networks (WSN), the IoT, wearable sensing and computing devices, AI and deep learning, smartphones and cellular technology, and cloud computing are some of such technologies. These technologies have witnessed marked advances in the recent past, and that trend is continuing. These advances can translate into drastic improvements for quality of life for the older adults and individuals living with dementia or disability. Thorough health and environmental monitoring, pervasive communication capabilities, intuitive interfaces, and powerful yet unobtrusive information collection, such technologies can provide transformative effects for aging and disability. While there have been many signs of progress in these areas and AT developments, many challenges also need to be addressed. These are discussed in this section.

Some samples of the AT and devices that provide solutions to issues encountered in aged care, dementia, or disability were examined in the previous sections of this chapter. Many beneficial types of equipment, as well as AT shortcomings, have obviously been left out. Nevertheless, the samples and discussions make three main points noticeable. Firstly, for various reasons, the adoption of the conventional and traditional AT is higher than many of the more sophisticated devices. Then, there are major AT improvements that have been made possible by adding the ever growing sensing capabilities. The third point relates to the high potentials for AT advances that can be realized by adding features provided by AI, deep learning, along with smart devices and environments. This last point can be extended to embed ambient intelligence and to include network and cloud-based AAL. These have been the subject of several studies, with comprehensive reviews published, for instance, see [55–57]. To set the scene for the next parts, a brief overview of some of these systems are outlined here.

Given the scope and the diversity of the areas included in AAL developments and implementations, many projects are part of some large initiative. For example, some of the projects are part of the Active and Healthy Aging of European Innovation Partnership (https://ec.europa.eu/eip/ageing/home_en). The AALIANCE2 is another example that is built upon extensive and fruitful projects aiming to provide AAL solutions for the well-being of older persons in Europe (http://www.aaliance2. eu/). It already has an extensive network of various stakeholders from many sectors in different countries. Some of the interesting outcomes of its parent project include business models and elaborate roadmaps for AAL developments with different strategies that can help older people remain active longer and can facilitate their social inclusion [58].

The use of technology in dementia and aged care is also necessary for addressing the shortages of formal caregivers and healthcare professionals, as well as the falling number of informal carers [18]. The latter problem, as discussed in Sect. 2.1, is exasperated due to an aging population and demographic changes. As is evident from Fig. 3.2, while the population is aging, the proportion of children who will act as the future carers is also dropping. Such demographic changes if not properly addressed can have devastating consequences. Part of the solution can be provided by advanced technologies, such as IoT-based systems and AAL. Obviously, the care for the aged people and those living with dementia remain a significant part of AAL. In these contexts, the European Commission has initiated the Ambient Assisted Living Joint Programme (http://www.aal-europe.eu/aal-joint-programme-event-at--the-eu-parliament/). Some samples of the many projects that are under this program are mentioned in this work. For all of these projects, the overarching principle is to enable a more independent life for older people, particularly those who may be socially isolated, or those living with disability or dementia.

FOOD is an example of a project aiming for the development of specific AAL services (http://www.food-aal.eu/). It is designed to assist with food preparation and related activities that support older adults in their daily living. It aims to promote the autonomy and independent living in all aspects of daily tasks, emphasizing kitchen-related activities making interactions with home appliances simple and safe. The developed solutions are based on the smooth incorporation of sensors, smart devices, and smart appliances, integrated with Internet-based services and applications. This combination can provide the needed food preparation functionalities and offer the

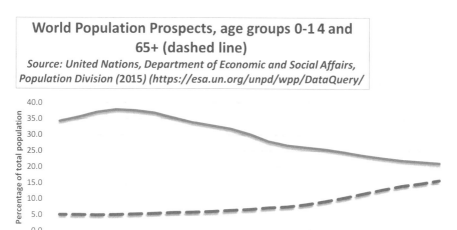

Fig. 3.2 Aging effect: Percentages of world population in two age groups (1950–2050)

needed information or communication capabilities through intuitive interfaces in the home or other supported environments. The network-based artificial and human intelligence can process and analyze the data from the sensors that are monitoring the people, appliances, and the environment, contributing to the support for independent living if required. The novelty of the project stems from its effective incorporation of the IoT, Semantic Web, and Web 2.0 in a cooperative manner.

The HOST is another specific project, which is part of the AAL (http://www. host-aal.eu). It addresses the needs of the elderly living in social housing establishments. The project aims to offer proper housing settings to the individuals who may have special needs and who cannot afford to rent accommodations that meet their needs. The required skills and other complexities associated with taking advantage of advances in technology can result in older adults, particularly those with lower incomes, not being able to benefit from ICT and digital society. The lack of awareness and capabilities to identify and use technologies may result in exclusions from the social life, which in turn can limit the independence of these individuals. In the application domain of the HOST, the clients are mostly seniors with low incomes. The project aims to provide easy-to-use technologies and services in these settings through implementing the "connected flats." Use of specific equipment to improve communication with family, service providers, and housing operators is believed to result in a more comfortable living for older adults and enhanc their social inclusion.

Another AAL project, NETCARITY, aims to demonstrate that the proper integration of technology into a familiar environment, like a home, can help with improving the social inclusion of the older adults living alone (www.netcarity.org/). In some sense, these are similar aims to those of the HOST project, but in different settings. NETCARITY uses touchscreen technology with easy to use interfaces, encouraging social interactions to reinforce ties with the family, friends, and the community of the elderly.

Some projects offer services for improving the quality of life for both the people living with dementia and their caregivers. MonAMI was one such project, carried out in Europe from 2006 under e-Inclusion topic (http://cordis.europa.eu). It was based on exploiting ambient intelligence technologies built on top of mainstream devices and services. Its aim was to minimize the exclusion risk for the elderly and individuals living with disability. It appears that its framework and use of sensors and actuators in a wireless environment has been of tangible benefit to people living alone with some cognitive impairment, like dementia.

Mylife is a more recent project for older persons with reduced cognitive abilities and to a lesser extent for supporting formal or informal caregivers (http://www.aal-europe.eu/projects/my-life/). It is based on providing multimedia services using simple, intuitive touchscreen interfaces to access services, such as listening to music, calendar, photo albums, watching video and other media that are available on the Internet. The interfaces of this AAL are customizable to address the individual's needs and desires [18].

There are obviously projects related to AAL for improving life quality for the seniors in other parts of the world. For example, Ubiquitous Computing and Network is a 10-year project supported by the government of the Republic of Korea. It is based on utilization of pervasive smart environments for life care. It aims to develop essential capabilities and technologies for wellbeing and safety of people who use it, based on community computing [59].

Some projects indirectly assist people with dementia, as their primary focus is on the caregivers. For instance, a daily life support system designed for use in aged-care facilities, aiming to alleviate the burden of the caregivers has been developed in Japan [60]. The system connects multi-purpose monitors, fall detectors, a terminal robot, and a PC through a local area network. Using the cameras and infrared sensors, the multi-purpose monitors allow for observing an individual's behavior and their potential wandering. It can monitor the living areas and the exits of the facility. Fall detectors can send alarms to caregivers. A terminal robot can help with promoting communication with caregivers, family, and relatives. This project, similar to many others, claims to have received positive feedback from the caregivers. However, most of such projects rarely appear to be complete and are in most cases can be considered as works in progress.

There are interesting, yet mostly incomplete, works that are based on using ambient intelligence to facilitate independent travel by the blind and visually impaired people. Some of these are based on replicating some of the functions of a guide dog [10]. The resulting AT needs to meet some general considerations, such as being light, unobtrusive, intuitive to use, and having proper interfaces. It also should be capable of real-time analysis of the surroundings to identify a suitable travel path avoiding any obstacles and be able to relay that information properly to the user. Many technology canes can detect obstacles and pass that information using audio and tactile interfaces to the cane bearer. They are mostly based on using a number of ultrasonic or laser light sensors, as previously discussed.

For navigation purposes, the assistive device must be capable of localization of the user, as well as pinpointing all objects that can be potential obstacles as the individual moves. It must also be able to generate the path for movement, based on the user intentions, and appropriately communicate the environment and navigation information to them. These are complex tasks that require the use of ambient intelligence and sophisticated strategies. In this area too, many systems have been proposed and developed, but many gaps and issues still exist. Most of them use a cane with augmented sensing capabilities, some communication and interface mechanisms, and local or remote computing services. These are usually complemented with GPS, cameras, digital maps, or some form of tags that form a guide.

The use of infrared LEDs that are placed strategically to act as a guide is not that new [61]. The information that they can provide is naturally limited by their number, location, and visibility by the receiver that is usually mounted on the cane. The receiver has to be small and lightweight, and consequently has a limited coverage area. Conversely, the LEDs can be installed on the cane, and their locations that will correspond to user's position and movements be deduced by using infrared cameras and other equipment [38]. This information plus a digital map describing the room can be used to calculate the paths to destinations and guide the user to avoid obstacles. The map needs to be refreshed if any object in the location is moved or added. The use of ambient intelligence for real-time recognition of indoor scenes for generation of dynamic maps may overcome parts of this issue [62]. Navigations that employ Simultaneous Localization and Mapping (SLAM) in dynamic environments methodologies and use smart environments, to be discussed in Sects. 5.1 and 6.1, can also address some of these issues [63].

For indoor navigation by the visually impaired people, RFID technology can also be used in fashions similar to the LEDS. The RFID tags that form a path can be followed by a blind person using a cane that carries an RFID reader [39]. However, from a practical point of view, it needs to be noted that RFID readers are usually bulky and too big for mounting on a cane. So, a more suitable but perhaps still impractical solution can be based on employing a simple robot carrying the reader that can follow the tags and guide the blind individual [64, 65]. Many of the works using RFID for indoor navigations, do not properly address the dynamic nature of objects and obstacle avoidance or steering in new and unknown environments.

As with using the LEDs to position the user, RFID tags can be worn by a blind person and placed on objects. These can then be read through room or ceiling mounted readers to identify them and track the user. Some works use this information along with the sensors of a smartphone that is equipped with an NFC interface, in conjunction with a semantic object-oriented model of the environment and various route-planning algorithms to provide voice-based directions to the user [41]. The tags need to be placed on all items of interest, like the doors, walls, and other objects, as well as on strategic points.

It can be constructive to note that companies like Google, IBM, Microsoft, and several others offer great solutions relevant to cognition. For example, Google CloudPlatform provides a service for analyzing natural spoken language and comprehending the overall sentiment that is expressed by it (https://cloud.google.

com/natural-language/). IBM Watson claims to be built on the power of cognitive computing, and its "Tone Analyzer Service can detect emotion, social tendencies, and language style from text" (https://www.ibm.com/cognitive/au-en/?#renderGridTiles). IBM Bluemix provides for a range of cognitive technologies that can be used in building smart applications for analyzing images and videos or understanding sentiments from the text (https://www.ibm.com/cloud-computing/bluemix/). Microsoft Cognitive Services "enables developers to easily add intelligent features – such as emotion and video detection; facial, speech and vision recognition; and speech and language understanding into their applications" (https://www.microsoft.com/cognitive-services/en-us/). While these services are not necessarily developed for providing intelligence and cognition capabilities for use in AT, they can be of value in the development of apps and technologies that are context-aware and responsive to the needs of any individual, including an aged person or one living with disability or dementia.

References

1. IEEE. (2016). IEEE Digital Senses. Available online, http://digitalsenses.ieee.org/ [Accessed 10-Oct-2016].
2. Shinohara, K., & Wobbrock, J. O. (2011). In the shadow of misperception: assistive technology use and social interactions. In *Proceedings of the SIGCHI Conference on Human Factors in Computing Systems* (pp. 705–714).
3. Cook, A. M., & Polgar, J. M. (2014). *Assistive technologies: Principles and practice*. Elsevier Health Sciences.
4. Garcia-Espinosa, E., Longoria-Gandara, O., Veloz-Guerrero, A., & Riva, G. G. (2015, October). Hearing aid devices for smart cities: A survey. In *Smart Cities Conference (ISC2), 2015 IEEE First International* (pp. 1–5).
5. Holube, I., & Hamacher, V. (2005). Hearing-aid technology. In *Communication Acoustics* (pp. 255–276). Springer Berlin Heidelberg.
6. Strom, K. E. (2006). The HR 2006 dispenser survey. *Hearing Review*, *13*(6), 16.
7. Wilson, B. S., & Dorman, M. F. (2008). Cochlear implants: a remarkable past and a brilliant future. *Hearing research*, *242*(1), 3–21.
8. Edwards, B. (2007). The future of hearing aid technology. *Trends in amplification*, *11*(1), 31–45.
9. Bledsoe, C. W. (1997). Originators of orientation and mobility training. Foundations of orientation and mobility, 580–623.
10. Hersh, M. A., & Johnson, M. A. (2008). Mobility: an overview. *Assistive technology for visually impaired and blind people*, 167–208. Springer London.
11. Stevens, J. C. (1992). Aging and Spatial Acuity of Touch. Journal of Gerontology 47 (1), 35–40.
12. Frank Lopresti, E., Mihailidis, A., & Kirsch, N. (2004). Assistive technology for cognitive rehabilitation: State of the art. *Neuropsychological rehabilitation*, *14*(1–2), 5–39.
13. Lancioni, G. E., & Singh, N. N. (2014). Assistive technologies for improving quality of life. In *Assistive Technologies for People with Diverse Abilities* (pp. 1–20). Springer New York.
14. May, M., & LaPierre, C. (2008). Accessible global positioning system (GPS) and related orientation technologies. *Assistive technology for visually impaired and blind people*, 261–288.
15. Bosco, A., & Lancioni, G. (2015). Assistive Technologies Promoting the Experience of Self for People with Alzheimer's Disease. *Rivista internazionale di Filosofia e Psicologia*, *6*(2), 406–416.

16. Ienca, M., Fabrice, J., Elger, B., Caon, M., Pappagallo, A. S., Kressig, R. W., & Wangmo, T. (2017). Intelligent Assistive Technology for Alzheimer's Disease and Other Dementias: A Systematic Review. *Journal of Alzheimer's Disease*, (Preprint), 1–40.

17. Spring, H., Rowe, M., & Kelly, A. (2009). Improving Caregivers' Well-Being by Using Technology to Manage Nighttime Activity in Persons with Dementia. *Research in Gerontological Nursing, 2*(1), 39–48.

18. Federici, S., Tiberio, L., & Scherer, M. J. (2014). Ambient Assistive Technology for People with Dementia: An Answer to the Epidemiologic Transition. *New Research on Assistive Technologies: Uses and Limitations. New York, NY: Nova Publishers*, 1–30.

19. Duff, P., & Dolphin, C. (2007). Cost-benefit analysis of assistive technology to support independence for people with dementia–Part 1: Development of a methodological approach to the ENABLE cost-benefit analysis. *Technology and Disability, 19*(2, 3), 73–78.

20. Duff, P., & Dolphin, C. (2007). Cost-benefit analysis of assistive technology to support independence for people with dementia–Part 2: Results from employing the ENABLE cost-benefit model in practice. *Technology and Disability, 19*(2, 3), 79–90.

21. van den Heuvel, E., Jowitt, F., & McIntyre, A. (2012). Awareness, requirements and barriers to use of Assistive Technology designed to enable independence of people suffering from Dementia (ATD). *Technology and Disability, 24*(2), 139–148.

22. Nisbett, R. E., Peng, K., Choi, I., & Norenzayan, A. (2001). Culture and systems of thought: Holistic versus analytic cognition. *Psychological Review, 108* (2), 291–310.

23. Hommel, B., & Colzato, L. S. (2015). Learning from history: the need for a synthetic approach to human cognition. *Frontiers in Psychology, 6*, 1435.

24. Halewood, A. M. K., Sabino, M., Sudan, R., & Yadunath, D. (2015). Six digital technologies to watch. *World Bank*.

25. Szolovits, P., Patil, R. S., & Schwartz, W. B. (1988). Artificial intelligence in medical diagnosis. *Annals of internal medicine, 108*(1), 80–87.

26. Nezhad, H. R. M. (2015). Cognitive assistance at work. In 2015 AAAI Fall Symposium Series.

27. Witten, I. H., Frank, E. A., Hall, M. J., & Pal, C. (2016). *Data Mining: Practical Machine Learning Tools and Techniques*. Elsevier.

28. Panou, M., Cabrera, M. F., Bekiaris, E., & Touliou, K. (2014). ICT services for prolonging independent living of the elderly with cognitive impairments-IN LIFE concept. *Studies in health technology and informatics, 217*, 659–663.

29. Li, R., Lu, B., & McDonald-Maier, K. D. (2015). Cognitive assisted living ambient system: a survey. *Digital Communications and Networks, 1*(4), 229–252.

30. Sharpanskykh, A., & Treur, J. (2012). An ambient agent architecture exploiting automated cognitive analysis. *Journal of Ambient Intelligence and Humanized Computing, 3*(3), 219–237.

31. Singh, N. N., Lancioni, G. E., Sigafoos, J., O'Reilly, M. F., & Winton, A. S. (2014). Assistive technology for people with Alzheimer's disease. In *Assistive technologies for people with diverse abilities* (pp. 219–250). Springer New York.

32. Hutchins, E. (1995). *Cognition in the wild*. Cambridge, Mass. MIT Press.

33. Alm, N., Dye, R., Gowans, G., Campbell, J., Astell, A., & Ellis, M. (2007). A communication support system for older people with dementia. *Computer, 40*(5), 35–41.

34. Norman, A. L. M. (2015). Distributed cognition, dementia, and technology. Assistive Technology: Building Bridges, 217, 319.

35. Csapó, Á., Wersényi, G., Nagy, H., & Stockman, T. (2015). A survey of assistive technologies and applications for blind users on mobile platforms: a review and foundation for research. *Journal on Multimodal User Interfaces, 9*(4), 275–286.

36. National Research Council. (1986). Electronic travel aids: New directions for research.

37. Vision Loss. (2016). Mobility Aids. Available online, http://visionloss.org.au/mobility-aids/ [Accessed: 10-Oct-2016]

38. Guerrero, L. A., Vasquez, F., & Ochoa, S. F. (2012). An indoor navigation system for the visually impaired. *Sensors, 12*(6), 8236–8258.

39. Na, J. (2006, July). The blind interactive guide system using RFID-based indoor positioning system. In *International Conference on Computers for Handicapped Persons* (pp. 1298–1305). Springer Berlin Heidelberg.
40. Di Giampaolo, E. (2010). A passive-RFID based indoor navigation system for visually impaired people. In Applied Sciences in Biomedical and Communication Technologies (ISABEL), 2010 3rd International Symposium on (pp. 1–5).
41. Ivanov, R. (2012). RSNAVI: an RFID-based context-aware indoor navigation system for the blind. In *Proceedings of the 13th international conference on computer systems and technologies* (pp. 313–320).
42. Kulyukin, V., Gharpure, C., Nicholson, J., & Osborne, G. (2006). Robot-assisted wayfinding for the visually impaired in structured indoor environments. Autonomous Robots, 21(1), 29–41.
43. Tsirmpas, C.; Rompas, A.; Fokou, O.; Koutsouris, D. (2015). An indoor navigation system for visually impaired and elderly people based on Radio Frequency Identification (RFID). Inf. Sci. 2015, 320, 288–305.
44. Yelamarthi, K., Haas, D., Nielsen, D., & Mothersell, S. (2010, August). RFID and GPS integrated navigation system for the visually impaired. In Circuits and Systems (MWSCAS), 2010 53rd IEEE International Midwest Symposium on (pp. 1149–1152).
45. Hernández-García, D. E., Gonzalez-Barbosa, J. J., Hurtado-Ramos, J. B., Ornelas-Rodríguez, F. J., Castaneda, E. C., Ramírez, A., & Avina- Cervantez, J. G. (2011, February). 3-D city models: Mapping approach using lidar technology. In *Electrical Communications and Computers (CONIELECOMP), 2011 21st International Conference on* (pp. 206–211).
46. Poulton, C., & Watts, M. (2016). MIT and DARPA pack lidar sensor onto single chip. *IEEE Spectrum.*
47. Watts, M. (2016). Photonic Microsystems Group News. Available online, http://www.rle.mit.edu/pmg/news/ [Accessed 10-Oct-2016].
48. Giudice, N.A. & Legge, G.E. (2008). Blind Navigation and the Role of Technology. In The Engineering Handbook of Smart Technology for Aging, Disability, and Independence; Helal, A., Mokhtari, M., Abdulrazak, B., Eds.; John Wiley & Sons: Hoboken, NJ, USA, 2008; pp. 479–500.
49. Nakajima, M.; Haruyama, S. (2013). New indoor navigation system for visually impaired people using visible light communication. EURASIP J. Wirel. Commun. Netw. 2013, 37..
50. Yavari, M., & Nickerson, B. G. (2014). Ultra wideband wireless positioning systems. *Dept. Faculty Comput. Sci., Univ. New Brunswick, Fredericton, NB, Canada, Tech. Rep. TR14-230.*
51. Cecilio, J.; Duarte, K.; Furtado, P. (2015). BlindeDroid: An information tracking system for real-time guiding of blind people. Procedia Comput. Sci. 2015, 52, 113–120.
52. Abe, Y., Toya, M., & Inoue, M. (2013). Early detection system considering types of dementia by behavior sensing. In *Consumer Electronics (GCCE), 2013 IEEE 2nd Global Conference on* (pp. 348–349).
53. Hamann, L. M. A., Airaldi, L. L., Molinas, M. E. B., Rujana, M., Torre, J., & Gramajo, S. (2015, December). Smart doorbell: An ICT solution to enhance inclusion of disabled people. In *ITU Kaleidoscope: Trust in the Information Society (K-2015), 2015.* (pp. 1–7).
54. Walker, S. (2016). Face recognition app taking Russia by storm may bring end to public anonymity. The Guardian. Available online, https://www.theguardian.com/technology/2016/may/17/findface-face-recognition-app-end-public-anonymity-vkontakte [Accessed: 10-Apr-2017].
55. Dohr, A., Modre-Opsrian, R., Drobics, M., Hayn, D., & Schreier, G. (2010). The Internet of Things for Ambient Assisted Living. *Information Technology: New Generations (ITNG), 2010 Seventh International Conference on,* 804–809.
56. Acampora, G., Cook, D. J., Rashidi, P. , & Vasilakos, A. V. (2013). A Survey on Ambient Intelligence in Healthcare, in *Proceedings of the IEEE*, vol. 101, no. 12, pp. 2470–2494, Dec. 2013.
57. Nakashima, H., Aghajan, Hamid K, & Augusto, Juan Carlos. (2010). *Handbook of ambient intelligence and smart environments.* New York: Springer.

58. van den Broek, G., Cavallo, F., & Wehrmann, C. (2010). *AALIANCE ambient assisted living roadmap* (Vol. 6). IOS press.

59. Kim, M., Cho, W. D., Lee, J., Park, R. W., Mukhtar, H., & Kim, K. H. (2010). Ubiquitous Korea Project. In *Handbook of Ambient Intelligence and Smart Environments* (pp. 1257–1283). Springer US.

60. Takahashi, Y., Kawai, T., & Komeda, T. (2014). Development of a Daily Life Support System for Elderly Persons with Dementia in the Care Facility. *Studies in health technology and informatics, 217*, 1036–1039.

61. Bentzen, B., Crandall, W., & Myers, L. (1999). Wayfinding system for transportation services: Remote infrared audible signage for transit stations, surface transit, and intersections. *Transportation Research Record: Journal of the Transportation Research Board*, (1671), 19–26.

62. Santofimia, M. J., Fahlman, S. E., del Toro, X., Moya, F., & Lopez, J. C. (2011). A semantic model for actions and events in ambient intelligence. *Engineering Applications of Artificial Intelligence, 24*(8), 1432–1445.

63. Davison, A. J., Reid, I. D., Molton, N. D., & Stasse, O. (2007). MonoSLAM: Real-time single camera SLAM. *IEEE transactions on pattern analysis and machine intelligence, 29*(6).

64. Kulyukin, V., Gharpure, C., Nicholson, J., & Pavithran, S. (2004). RFID in robot-assisted indoor navigation for the visually impaired. *2004 IEEE/RSJ International Conference on Intelligent Robots and Systems (IROS), 2*, 1979–1984.

65. Tsukiyama, T. (2003). Navigation system for mobile robots using RFID tags. In *Proceedings of the International Conference on Advanced Robotics (ICAR)*.

Chapter 4
The IoT and Smart Environments:
An Overview

The enterprises are beginning to take advantage of the opportunities that the IoT offers by implanting sensors and actuators into their products and services. The IoT adoptions have also enabled the development of smart environments. These advances are drastically shifting the way businesses function and how people engage with the physical world. The pervasive nature of the IoT and the context-aware characteristics that stem from connecting everything together, support the provisions of ambient intelligence [1]. The capabilities to communicate such intelligence with other machines, including those nearby and those providing cloud-based deep learning and AI capacities, hold significant potentials for improving people's lives in many respects. This chapter outlines the fundamental structure and characteristics of the IoT and smart environments, focusing on those features that can be substantial for assistive technology developments and employments. The chapter also presents some of the typical applications of the IoT and smart environments, to demonstrate the advantages and implications of their deployment in aged-care and for empowering people living with disability or dementia.

4.1 The IoT: Sensors, Actuators, and Communications

A 2015 review of More than 20,000 scholarly documents published since 1995 has identified 14 concepts that shape the future of information infrastructures [2]. The dominant concepts were identified to be semantic web, ubiquitous and pervasive computing, and to a lesser extent ambient intelligence and smart environments. All of these concepts come together in the IoT. The IoT is actually shaping the future of various permutations of interactions between machines and humans. Its magnitude, applications, and value are expanding rapidly. Many of the existing devices and services can be easily integrated with the IoT. People can also be equipped with many sensors to track their health-related attributes, presence, position, movements, or even sentiments and attitudes.

© Springer International Publishing AG 2017
S. Shahrestani, *Internet of Things and Smart Environments*,
DOI 10.1007/978-3-319-60164-9_4

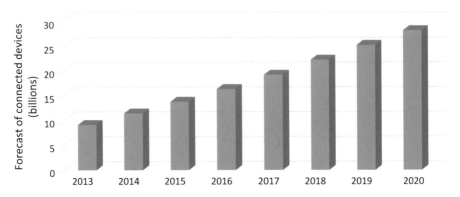

Fig. 4.1 Forecast of Connected devices in billions. Data Source: [4]

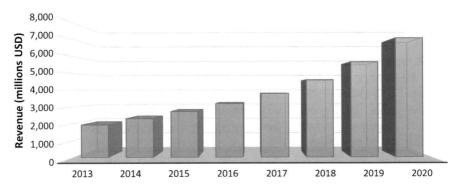

Fig. 4.2 The IoT Revenue Forecast. Data Source: [4]

The IoT is projected to globally generate anywhere from US$2.7 trillion to the US$14.4 trillion in value by 2025 [3]. These projections, forecast that in the healthcare sector alone, the productivity gains and cost savings can amount to US$1.1 trillion to US$2.5 trillion in value. As depicted in Fig. 1.1 in Sect. 1.2, industry forecasts for the number of connected devices varies between 25 and 50 billion by 2020. AT&T and IDC analysts have provided some of the most conservative forecasts for connected devices displayed in Fig. 4.1 and for generated revenue shown in Fig. 4.2.

Even with such magnitudes, there is no clear and universally accepted definition for the evolving IoT, and for that matter, for things. Many definitions have been suggested that contain common words like "things, interconnected, the Internet, cyber-physical, devices, sensors, smart, physical objects, and virtual world" [5]. The World Bank considers the IoT as the interconnection of objects to the Internet "through embedded computing devices, such as RFID chips and sensors. IoT products can be classified into five broad categories: wearable devices, smart homes, smart cities, environmental sensors, and business applications…" [6]. The first four of these categories are directly relevant to the theme of this book. The International

Telecommunications Union (ITU) defines IoT as, "A global infrastructure for the information society, enabling advanced services by interconnecting physical and virtual things based on existing and evolving interoperable information and communication technologies" [7]. For this work, these definitions provide sufficient clarity.

The IoT is an infrastructure that is built around physical objects, sensors, and actuators that have data communication capabilities. This infrastructure can furnish for environmental monitoring, tracking of entities, measuring of parameters, or controlling systems over a network or the Internet. The range of things and devices that make up the IoT is vast. It usually contains small and inexpensive objects, such as RFID tags that cover a broad range by themselves, commercially costing anything between 10 cents to US$50 (http://www.barcoding.com/). Expensive and highly sophisticated devices like smartphones can also be part of this infrastructure. However, in the lack of a universally accepted definition for the IoT, significant variations of the naming of devices, virtual objects, and services do occur.

The IoT involves two major characteristics, neither of which are that new. The first one relates to objects or things; they all have or can have sensing or actuation capabilities, and some may also have other data processing or retention capabilities. The second one is the capability of direct M2M communication and in some cases, control over the Internet. The IoT revolution relates to the prospects for building systems and smart environments that interconnect a massive number of devices that can be quite diverse with a large range of capabilities and applications.

The capacity to communicate the presence and identity of one thing with another thing can be used in automation of some applications or processes. For instance, when a particular smartphone gets close to a door, the door can unlock or lock automatically, or a notification can be sent to another device. The sensor networks can measure many attributes of the physical world around them continually and communicate them with other sensors and services. In some ways, the human sensory systems of sight, hearing, smell, taste, and touch may be inferior to the sensors of the IoT [8]. In such cases, the IoT can complement human senses and perception by providing additional information through, say a smartphone. Of course, there are many aspects of human sensory and cognition systems that the IoT inferiority is evident. This weakness is particularly the case for conveying the information to a user in efficient manners, as well as the coordination between the many sensory and actuation entities that can be involved in even simple everyday activities of human beings. The machine-human infusion, while not yet a reality, is moving forward quickly.

The increasing dependence of people with all types of abilities on sensors and AI-empowered systems that talk and behave like humans can result in smart things and environments intermediating and augmenting our senses. For instance, a new machine-brain interface company, Neuralink, has been founded with the aim of using what is called "neural lace" to bridge machines into human brains [9]. When and if realized, this goes well beyond human-machine infusion, as machines can replace parts of human sensory and cognition functions in an integrated fashion. While these works do not appear to be pursued for direct benefit to the elderly or people living with disability, their assistances for these individuals should be obvious. It is interesting to note that the idea behind the machine-brain interface transpired in 2011 aiming "for an electronic brain chip to treat traumatic brain injury... to reestablish damaged connections by recording neurons in one part of the brain, then transmitting the chatter to another..." with a successful prototype demonstrated to help brain-damaged rats by 2013 [10].

Smart environments, homes, and cities can now realistically capitalize on devices and products that embed sensors or actuators and have wireless communication capabilities to facilitate carrying out tasks that require intelligence, without direct human involvement. Many everyday appliances and machines are already network-connected, with sensing and other capabilities that make them considered as smart devices. The range of these appliances and the diversity of tasks that they can automatically carry out are rapidly increasing. The European Commission considers a smart city a place that digital and communication technologies are used to improve the provided services for the benefit of its residents (https://ec.europa.eu/digital-single-market/en/smart-cities). This definition can translate into "smarter urban transport networks, upgraded water supply and waste disposal facilities, and more efficient ways to light and heat buildings." By some accounts, there were 21 smart cities across the globe in 2013, forecast to rise to 88 or more by 2025 [11].

Miniaturization, improved sources of power, better energy management, enhanced AI techniques, and huge advances in communication technologies have all contributed to the use of very small sensors in many physical devices and environments. These small, low cost and low power things can run for extended periods of time without much attention from human beings. The sales of sensors have seen annual growths of 70% since 2010 [12]. A smartphone that typically had six sensors in 2010 was equipped with 16 sensors in 2014, and the newer ones have 20 sensors or more, including proximity, magnetometer, touch, GPS, position, and gyroscope. Adding actuators can further facilitate many autonomous systems that can achieve various functions, requiring only skeletal interventions from humans.

Wearables are among the fastest growing parts of the IoT technologies. These portable accessories include clothes, glasses, and watches as perhaps the most prevalent one. They are widely accepted for fitness and activity tracking, which are based on collecting data that is communicated with other devices, like the user's smartphone or some servers. Sophisticated smartwatches can act as networked communication devices with many sensors that unlike most other devices, can remain in constant physical contact with their users. Their market is predicted to grow to 214 million units by 2018 [13]. Other analysts have forecast that the worldwide wearable shipments will reach 420 million by 2020, more than a five-fold increase from the 80 million shipped in 2015 [14]. Other parts of the IoT-related technology also enjoy rapid growth. Figure 4.3 presents some indicators of that growth between the years of 2013 and 2020, using data from various industry forecasters that have been previously summarized [3].

The huge number of connected objects, diverse applications, and the market share of the IoT comes with several challenges. Perhaps, the most prominent ones are those related to security and privacy issues. The low processing power of the

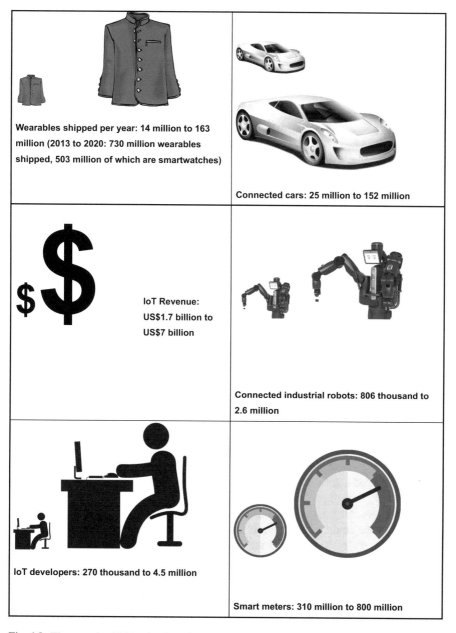

Fig. 4.3 The growth of IoT technology from 2013 to 2020. Data Source: [3]

things, arising from their main advantages of being low-power and small, can also limit their capabilities for using many security services and encryption methods [15]. Poorly secured IoT devices can be targeted, causing them to malfunction, or can be used for cyber attacks on other systems. An example of this limitation is the well-publicized narrative on fridges mounting Distributed Denial of Service (DDoS) attacks [16]. The things can cause serious privacy concerns for their users or owners. The popular fitness tracking devices and many other objects of the IoT are built around machine-to-machine communications and processes that tend to operate automatically, without human involvement and awareness. As such, they may not even alert the user that they are collecting data about the individual and potentially sharing it with third parties. Things can expose user data and information and may have no clear way or protocol to warn the user when a security problem arises. Many organizations, including ITU, have published several recommendations on security and privacy of the IoT, things, and their owners or users, for example, see [17, 18]. Some of these challenges will be further explored in Section 5.2.

4.2   Smart Buildings, Cities, and Environments

The communication and sensing or actuation capabilities of smart things can be used to get and transmit large amounts of rich information from many devices, regarding the activities, behavior, and environmental parameters that can be of interest. As discussed previously, this enriched data can be analyzed locally, on devices like smartphones, or remotely, on cloud-based deep learning and AI services. The analysis can, in turn, facilitate the discovery of relevant insights regarding the behavior and needs of the residents, the services, the organization, utilities, and the infrastructures of the home or city. Such environments, as we have seen, can find many applications in projects for incorporation of the IoT for AAL [19, 20]. More generally, they enable what can be referred to as smart homes, buildings, or cities.

A smart city employs ICT for continued development of the environment and for improving the quality of life of its citizens with public services that are interactive, efficient, and accessible [21]. Smart cities collect, integrate, and analyze data, as systemic as practicable, to intelligently reduce pollution, improve waste removal and recycling, manage traffic congestions, tackle poverty and homelessness, improve accessibility and inclusion for all individuals. Smart cities are fueled by many IoT-based smart devices, such as smart meters or lights. They also use IoT-leveraged smart structures, like smart parking or transportation systems.

The global urbanization is one of the main drivers for the smart city concepts and visions. For the developing world, the World Bank forecasts a doubling of the number of people living in cities between 2000 and 2030, adding two billion new city inhabitants [22]. The sheer number of new city dwellers makes it imperative to use technology for better decision-making, policy development, improvement of services and organizations, or even enforcement of regulations. The same Report from the World Bank indicates that many governments have already started responding to this need. Rio de Janeiro in Brazil has implemented a novel approach to bringing data from over 30 agencies and services in an Operations Center, resembling the US National Aeronautics and Space Administration. China has provided US$70 billion smart city line of credit, with an investment fund of US$8 billion. India is pursuing

Libelium Smart World

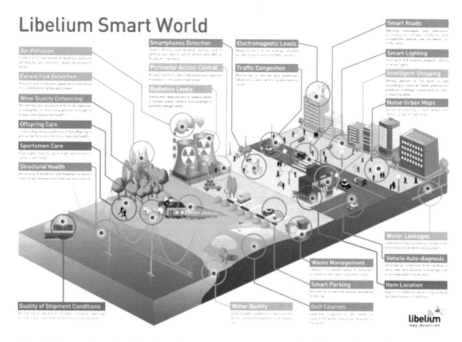

Fig. 4.4 A sample of applications in smart environments, used with permission from Libelium (http://www.libelium.com/libelium-smart-world-infographic-smart-cities-internet-of-things/)

the building of 100 smart cities. Overall, it appears that Asian countries are the global market leaders in embracing smart cities and their related projects.

Many smart city projects in EU and North America aim to improve the efficiencies in existing infrastructures, services, and in the daily lives of people. They mostly aim to overcome the resource constraints and to mitigate the urbanization effects on environmental sustainability. For example, they use the IoT technologies to decrease fossil fuel use to lower emissions or to reduce road congestions [23].

The starting point for many advances towards the smart cities are the massive data exchanges, through cellular systems and smartphones or social media. Theses convenient information exchanges allow for sharing of ideas and data amongst individuals, organization, and government agencies paving the way for identifying the required changes and acting upon them. They clearly fit the main criteria in most EU projects on smart cities, which can be summarized as, translation of technology into improved public services for citizens, sustainable use of resources, and less environmental impact (https://ec.europa.eu/digital-single-market/en/smart-cities). The concepts of smart cities include, but can go well beyond, the crowdsourcing ideas. Many interesting applications have been part of the concept of smart environments. Figure 4.4 presents a sample of 50 sensor based applications for smart environments from an infographic by a private sector company, Libelium (http://www.libelium.com/).

The inclusion of the vast number of IoT devices that facilitate typical applications for a smart city has been widely supported by the European Council's Framework Seven Programme (FP7). The relevant projects from that Programme continue to be funded and pursued by HORIZON 2020 (http://ec.europa.eu/programmes/horizon2020/). These Programmes have hosted many projects like Smart Santander in the Spanish city Santander, Sense Smart City in Sweden, and Amsterdam Smart City in Holland, among several others [24]. The Smart Santander facility has embedded thousands of various IoT nodes, mostly with Wireless Personal Area Networks (WPAN), RFID, and cellular system communication capabilities (http://www.smartsantander.eu). These include 400 parking sensors, 2000 RFID tags to smartly label points or places of interest, 50 irrigation monitors, and around 2000 sensors for environmental parameters, such as temperature, noise, ambient light, and air quality.

The deployment of things and the IoT in buildings and homes can also assist their residents with automation, delivery of more adaptive and individualized services, and better communication capabilities. The explosive growth of the IoT has made it possible for these not-so-new visions to be realized. Gartner is forecasting that by 2022, a typical family home can incorporate more than 500 smart devices [25]. Smart homes can monitor the movements, habits, and behavior of their dwellers to deploy suitable actuators, for example, to turn the lights and air conditioners on or off, to turn TV on or off, to lock and unlock doors, or to alert others [26]. In this fashion, the smart home as a daily living space can significantly improve the quality of life for people [27]. In a typical smart home sensors and actuators that are capable of communicating with other devices and people are in control of many appliances, doors, switches, and security of the space. The capability of such an environment to adapt to the daily living activities, habits, and abilities of a resident, makes a smart home particularly of value for independent living by older adults, or individuals with disability or dementia [28].

The smart and enabling environments require multi-disciplinary work and research that incorporates various concepts and technologies. These include sensor technology, universal design and accessibility, user profiling, context awareness, AI and deep learning, computer and communication technologies, human-machine interactions and interfaces, and AT [24]. The IoT can address many of the propositions for an enabling and empowering setting to provide individualized services for improving the inclusion. The IoT infrastructure in a daily living environment can be practically implemented to provide ambient intelligence. In this context, a smart home can shift the focus away from the assistive devices or technology towards objects or things, putting the emphasis on how the environment can accommodate an individual's daily living requirements in the set-ups that they can construct for themselves. In such a situation, the technology can be truly integrated and not be intrusive. It can be empowering and part of a holistic approach to improving the quality of life for its users.

Smart homes have developed rapidly into mainstream dwellings. However, to increase their adoption by the general public, as well as employment in AAL applications, some challenges need to be addressed. Perhaps, for many people the risks or the perceptions of having their privacy violated, prevents them from fully embracing the smart homes [29]. Other barriers can stem from feeling threatened by the technology and its disruptions to existing familiar care and services, lack of convincing evidence, funding issues, unfavorable policy decisions, and lack of information and awareness [30].

4.3   Typical Applications of the IoT and Smart Environments

The potential or realized applications of the IoT cover many diverse areas, which are widely published. For some representative examples, see [31–34]. The applications range from many smart and somehow overblown designs of objects to highly utilized clinical and health-related areas to manufacturing, agriculture, autonomous vehicles, among many others. According to McKinsey Global Institute, in 2025 the largest areas of IoT applications are forecast to be health, followed closely by manufacturing [35]. Based on that data, Fig. 4.5 shows that these two areas, healthcare and manufacturing, are projected to constitute more than three-quarters of the of the huge IoT market.

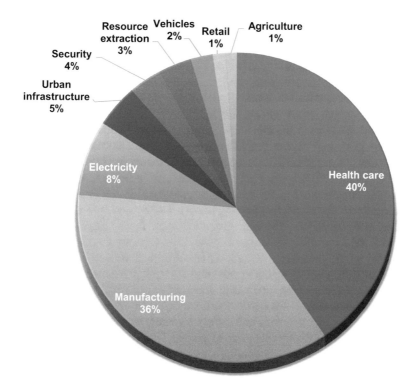

Fig. 4.5 Forecast of IoT Applications Market Share in 2025. Data Source: [12]

The extensive capabilities of the IoT in monitoring and automation, supplemented by the power of data analytics coming from deep learning and AI, can identify inefficiencies quickly and in many cases suggest remedies to overcome them. These can result in significant economic benefits for individuals, as well as for public and private sectors. Smart buildings and cities can reduce pollutions, improve services, enhance safety and security, and make life more inclusive for all their dwellers. The smart monitoring capacities of the IoT can also be used to promote and accomplish preventative health solutions to assist with early detection of some diseases, to help patients with keeping up with their treatments and supporting caregivers and healthcare professionals. The biggest share of the IoT market is forecast to be healthcare, which is also related to the theme of this work. Typical health-related applications will be discussed in the next section. In the remainder of this section, a small sample of other application areas will be presented.

The IoT has the potential of significantly improving the efficiency of operations and processes in manufacturing. Extensive uses of sensors can provide real-time monitoring of machinery and equipment, improving maintenance, reducing downtimes, and predicting needed repairs to avoid expensive overhauls or substitutions due to unnoticed faults. The use of RFID tags for inventory checks and tracking or management of supply chain material and products is already well established. Adding to these capabilities with additional IoT devices can provide more useful information for automating inventory management, for instance. They can improve synchronization between different parts of the manufacturing processes, to reduce delays, to make the flow transparent, and to reduce the required inventories to reach self-optimizing production.

Smart grids, transparent distributed generation, and optimal demand management are among the other obvious areas of IoT applications that can result in significant energy market improvements. By facilitating real-time machine-to-machine communication between the consuming equipment or appliances and the supply side, smart grids and meters provide many advantages. For example, the power company devices can inform the consumer side, of the applicable tariffs in real-time and let them choose to stay on or turn themselves off. The utility companies can also be given the authority to turn off or on some pre-specified equipment or appliances automatically. With the IoT, these arrangements can be made for one or all equipped machines and appliances. For the generation side, these smart interactions can reduce the need for building extensive generation capacity that is kept in reserve, just to meet the demand during the short periods of peak usage.

Urban infrastructures, including transport systems, traffic management, public services, safety and security, and waste management can also benefit significantly from the implementation of the IoT. Sensors can, for instance, monitor traffic congestions and communicate with traffic light actuators or suggest alternative routes to cars or buses. By utilizing such a smart system, London, Singapore, and Houston have significantly reduced commuting times for their residents [12]. The delay reductions will significantly improve productivity, reduce waste of energy, and improve air quality.

Some benefits and applications of smart buildings and cities have already been discussed in the previous section. With consumer enthusiasm and very obvious benefits, connected cars and autonomous vehicles have seen leaps of progress in the recent past years. The reasons for adding sensors to cars include lessening the number of crashes and accidents, fuel savings, congestion prevention, and productivity gains. It is forecast that after their full market deployment, self-driving cars will save the world economy US$6.9 trillion per annum [36]. This forecast for the US alone is $1.3 trillion per year, comprising of $656 billion in many productivity gains, $488 billion gains from prevented accidents, and $158 billion of savings in fuel costs. For usages of the IoT in smart homes, consumers have predominantly opted for devices and services that are related to safety and security. While in 2014 self-adjusting smart thermostats were the most popular products, in 2015 security was the main reason for 90% of US consumers to purchase a system related to smart homes [25].

Waste management may sound like an inexplicit area for the IoT application. However, several novel usages have been explored. A European Commission technical study has shown that the use of RFID in the recycling industry can reduce waste collection costs by up to 40% [37]. An interesting approach to waste management in the cities of Cleveland and Cincinnati in the US demonstrates some of the tangible benefits of using the IoT and RFID tags. In those projects, trash and recycling cans are equipped with RFID tags associated with the residents and their addresses. As the cans are tipped into the collecting truck, readers scan the tags. This data is analyzed for various purposes. It can indicate the time it takes for drivers to move from one house to another, helping to keep productivity up. The tags can help the crew to see which houses are recycling or putting the excessive rubbish out. This observation is considered to be important, as based on a recent analysis, 42% of the Cleveland trash collected by the city is recyclable, sellable for US$5.5 million (http://www.waste360.com/Collections_And_Transfer/rfid-waste-collection-201010). As a result of these initiatives, recycling volume in Cincinnati grew by 49%, and the residential waste volume fell by 17%. Cleveland has cut the related operating cost by 13% [12].

This section can by no means provide an exhaustive list of the various IoT application areas. That is well beyond the scope of this work. The application areas considered in this section can only serve as an incomplete sample, showing the diversity and the massive growth of the IoT. Some of them can also provide the ground-breaking ideas that can be of value to the older adults or individuals living with disability or dementia. The next section gives a sample of typical health-related applications of the IoT.

4.4 Typical Health-Related Applications of the IoT and Smart Environments

Healthcare is among the most appealing areas for the IoT application. As Fig. 4.5 in the previous section indicates, it is forecast to be the largest IoT application area. The IoT can expand and improve many health-related systems, such as remote health monitoring schemes that are based on using the Internet. The comprehensive monitoring and communication capabilities of the IoT can also be beneficial in applications relevant to fitness programs, aged care, warning of emergency situations, in-hospital patient observations, access to real-time patient data that is collected without intrusive devices, and even fighting counterfeit medications and drugs. They can overhaul the management of treatments of patients with chronic conditions, by providing in-time information and monitoring compliance while the patients are outside the direct care of the healthcare professionals. This section provides samples of some of the typical health-related IoT applications, with emphasis on those relevant to AT and the themes of this work. More examples of the IoT-based healthcare applications can be found in [38–40], and references therein.

As with the general applications, the improvements in efficiency and making novel approaches possible, are the main drivers for deploying the IoT in healthcare. In many situations, such deployments can improve the quality of life for the individuals and enrich their experience, while reducing the costs to them and the society. For instance, McKinsey puts the costs of treatment for chronic diseases at approximately 60% of total healthcare spending, with the annual global cost of treatments reaching US$15.5 trillion in 2025 [12]. They estimate that remote monitoring can reduce this cost by 10–20%. The savings relate to efficiencies from monitoring patients at home and perhaps more frequently, saving travel times and expenses, or more importantly being alerted to problems more quickly, avoiding potential emergencies and costly hospitalizations.

Generally speaking, the IoT-based healthcare may utilize a combination of medical devices, some form of wireless sensor networks, wearable devices, tags, mobile computing devices, and higher level communication and interface devices, such as smartphones. A good summary of various healthcare applications and their associated sensors and operations are given in Table 1 of the survey paper [38]. In most applications, the interactions among different things may employ one or several short-range communication approaches that use WPAN and WBAN protocols including, 6LoWPAN, RFID, NFC, BLE, Wireless Local Area Network (WLAN), and 802.11ah extension. Long range systems, like cellular systems or other broadband connections, are also usually required.

Smartphones, with their pervasive availability, can be a major component of the IoT-based healthcare applications. They can provide the capabilities for long range communications, act as an interface between things and humans, and contain suitable apps. A large and rapidly growing number of apps have been developed for healthcare. An excellent survey of these apps, not specific to the IoT, is given in [41, 42], and Table 2 in [38].

The demands of the aging population have resulted in the rapid deployment of some applications relevant to smart home technologies. These technologies can be of benefit to the elderly in terms of their independence, inclusion, and overall quality of life. In some situations, they can be of even more value to their caregivers. In some surveys, 50% of all consumers, 72% of those aged 25–34, and 74% of parents have said that their worries will be seriously alleviated if their parents or grandparents had smart home technology [25]. This attitude is important to note for its potential effects on many human lives as well as the opportunities it identifies for the smart home industry. The reasons behind such worry alleviations are related to the various benefits of the IoT and smart environments that are discussed in many parts of this work. One specific aspect that is worth mentioning again here relates to their use in early detection of dementia that was referred to in Sect. 2.3.

Early detection of dementia can help with slowing down its progress. The early signs of the onset of the disease are usually detected by those living with the affected individual. For those at risk, for example, based on their age, and particularly for those living alone, the IoT and smart homes can be of great benefit for early detection of dementia symptoms. Such a detection can be based on collecting data from the sensors in the home of the elderly and analyzing them in the cloud, for any signs of suspected cases. The information can then be used to alert the individuals, their carers, or healthcare professionals for follow-ups and possible diagnosis. An example of this approach is detailed in [43]. The efficiency and suitability of such an approach depend on many complex factors. For example, on the number of sensors, their types, and where they are located, acceptability by the user, costs, analytical capabilities and their precision, and their capacities to account for individual variations. These factors make it important to choose the minimal number of attributes that can automate the uncovering of the suspicion of dementia. A system that is based on these notions is reported in [44]. The proposed system accumulates many attributes of the users and employs feature extraction, threshold-triggered deviation monitoring, and pattern recognition to automate the detection process. It uses ZigBee and a Raspberry Pi for its M2M gateway. The sensors detecting human presence can trigger other sensors, for instance, to check for the sound of water running from a tap or to probe if the TV is on or off. Based on a combination of evaluation criteria, for example, forgetting to turn off a faucet, to take a shower, or wandering and its time in relation to sleep patterns, among others, the system claims to have been able to detect early signs of dementia with an accuracy rate of between 80% and 100%.

The elderly or any others with cognitive impairment can also benefit from smarter AT for wayfinding, identifying a path from where they are to their destination. The wayfinding issues can be simply age-related or stem from more severe cognitive decline or impairments of dementia. These problems can result in anxiety and dependence on others even for activities that the individual could normally carry out without any difficulties. Wayfinding, orientation, and spatial cognition depend on the individual's sensory and perception abilities, information-processing capabilities, previously learned knowledge and access to it, and motor capabilities [45]. All or some these four capabilities can be limited for individuals with a sensory or cognitive impairment or the older adults. As such, there are many applications of the IoT and smart environments that deal with the wayfinding issues. These can be related to diverse groups of people, for example, aiming to serve the blind and visually impaired people or being meant for employment by the aged persons with dementia.

Various forms of technologies have been used for wayfinding. RFID, Bluetooth, cellular technology, Wi-Fi, and talking signs are just some of them. In most of these solutions, the user needs to run applications on a mobile device, such a smartphone and be able to negotiate the use of software and at least some aspects of these technologies. The requirements for such abilities can be a critically negative point for some of the end users, particularly for those with cognitive impairments or for those unfamiliar with the required devices, and for the elderly, in general [46]. Talking signs can overcome some of these issues. They provide a repetitive, directional voice message, originating from the sign, transmitted to mobile devices by infrared (IR) signals. The directional characteristics of the IR beam mean that the signal strength increases as the user carrying the mobile device moves in the right direction towards the talking sign [47]. However, such a directed IR transmission requires a line of sight from the sign to the device, severely limiting the general deployment of the talking signs.

Wearables, and particularly smartwatches, can be of significant value in the IoT-based healthcare and wellbeing. Activity and fitness trackers already enjoy a big market, as discussed in the previous section. With the advances in sensor technologies, wearables can make more accurate measurements of the biological parameters. Advances in communication and analytical capabilities mean, making good use of this vast amount of biological data that can correspond to an individual's health, is readily achievable. With a direct connection to the skin, smartwatches facilitate convenient sensing of heart rate and its variability, temperature, blood oxygen content, and Galvanic skin response that can be used to identify physiological arousal and emotions [13]. Analysis of this data can be combined with other techniques, such as those recognizing facial expressions, to ascertain the emotional state and health of an individual which can, in turn, be used for diagnoses of some diseases, such as schizophrenia, depression, and autism [48].

The IoT and wearables are also providing many rather unconventional applications relevant to health and well-being. For instance, MIT in partnership with Microsoft has introduced DuoSkin that can be used to control mobile devices, and display or store information on skin, while resembling jewelry-like temporary tattoos (http://duoskin.media.mit.edu/). DuoSkin has demonstrated to be capable of three types of on-skin interfaces: sensing touch input, displaying output, and wireless communication [49].

Another example is that of La Roche-Posay, the parent company L'Oréal who has released the first stretchable electronic for consumers, My UV Patch (http://www.laroche-posay.com.au/article/Save-Your-Skin-UV-Patch/a27999.aspx). The patch is thinner than a normal Band-Aid, sticking to skin for up to 5 days, it uses connectable sensors to measure skin exposure to ultraviolet light. Using NFC, it can communicate with an app on a smartphone, allowing a user to monitor their UV exposure. The feasibility of using similar wearing sensor technological concepts to monitor Parkinson's Disease patients in different settings have also been trialed, with results yet to be published (http://www.businesswire.com/news/home/20170105005462/en).

The take-up of the health-related applications of the IoT and smart environments face a number of challenges. The dominant one relates to user privacy and data security. Most of these applications monitor sensitive information about their users and their environments. The users' concerns will only exacerbate with the increased deployments of the IoT, resulting in higher levels of shared data. Security and privacy requirements for the IoT-based healthcare solutions are similar to those for other medical and communication services [38]. Some of those requirements and their related standards were mentioned in Sect. 4.1. Another challenge arises from the proprietary nature of many medical devices, making them incapable of convenient communication with other machines to share data (http://medicalinteroperability.org/internet-of-things-impacts-hospitals-health-care-facilities/). These challenges for some of the IoT-based deployment scenarios, along with many of the promising ways to address them, will be discussed in more detail in the next chapter.

References

1. Dohr, A., Modre-Opsrian, R., Drobics, M., Hayn, D., & Schreier, G. (2010). The Internet of Things for Ambient Assisted Living. *Information Technology: New Generations (ITNG), 2010 Seventh International Conference on,* 804–809.
2. Olson, N., Nolin, J., & Nelhans, G. (2015). Semantic web, ubiquitous computing, or internet of things? A macro-analysis of scholarly publications. *Journal of Documentation, 71*(5), 884–916.
3. Castillo, A., & Thierer, A. D. (2015). Projecting the growth and economic impact of the internet of things. Available online: https://www.mercatus.org/system/files/IoT-EP-v3.pdf [Accessed 13 April 2017].
4. Lund, D., MacGillivray, C., Turner, V., & Morales, M. (2014). Worldwide and regional internet of things (iot) 2014–2020 forecast: A virtuous circle of proven value and demand. *International Data Corporation (IDC), Tech. Rep.*
5. Voas, J. (2016). Demystifying the Internet of Things. *Computer, 49*(6), 80–83. ComputerJune2016IoTDemystify
6. Halewood, A. M. K., Sabino, M., Sudan, R., & Yadunath, D. (2015). Six digital technologies to watch. *World Bank.*
7. International Telecommunication Union (2012). Overview of the Internet of things. ITU. Availabe online, https://www.itu.int/rec/T-RECY. 2060-201206-I [Accessed 10-Oct-2016].
8. Nolin, J., & Olson, N. (2016). The Internet of Things and convenience. *Internet Research, 26*(2), 360–376.
9. Weinberger, M. (2017). The smartphone is eventually going to die, and then things are going to get really crazy. Business Insider. Available online, https://www.businessinsider.com/death-of-the-smartphone-and-what-comes-after-2017-3#x3FEEWK9yVtezYd0.99 [Accessed 13 April 2017].
10. Regalado, A. (2017). Meet the Guys Who Sold "Neuralink" to Elon Musk without Even Realizing It. MIT Technology Review. Available online, https://www.technologyreview.com/s/604037/meet-the-guys-who-sold-neuralink-to-elon-musk-without-even-realizing-it/ [Accessed 13 April 2017].
11. HIS (2014). Smart *Cities to Rise Fourfold in Number from 2013 to 2025*; IHS Inc. Available online, http://news.ihsmarkit.com/pressrelease/ design-supply-chain-media/smart-cities-rise-fourfold-number-2013-2025 [Accessed 10-Oct-2016].
12. Manyika, J., Chui, M., Bughin, J., Dobbs, R., Bisson, P., & Marrs, A. (2013). *Disruptive technologies: Advances that will transform life, business, and the global economy* (Vol. 180). San Francisco, CA: McKinsey Global Institute.
13. Rawassizadeh, R., Price, B. A., & Petre, M. (2015). Wearables: Has the age of smartwatches finally arrived? *Communications of the ACM, 58*(1), 45–47.

14. Bootan, J. (2016). How I became a cyborg and joined an underground medical movement. MarketWatch. Available online, http://www.marketwatch.com/story/i-joined-an-underground-medical-movement-but-had-to-become-a-cyborg-to-do-it-2016-11-15 [Accessed 10-Jan-2017].

15. Chaudhuri, A. (2016). Internet of things data protection and privacy in the era of the General Data Protection Regulation. *Journal of Data Protection & Privacy, 1*(1), 64–75.

16. Grau, A. (2015). Can you trust your fridge?. *IEEE Spectrum, 52*(3), 50–56.

17. International Telecommunication Union (2014). Common requirements for Internet of things (IoT) applications. ITU-T F.748.0. Available online, https://www.itu.int/rec/T-REC-F.748.0-201410-I/en [Accessed 10-Oct-2016].

18. International Telecommunication Union (2014). Common requirements of the Internet of things. ITU-T Y.4100/Y.2066. Available online, https://www.itu.int/rec/T-REC-Y.2066-201406-I/en [Accessed 10-Oct-2016].

19. Domingo, M. C. (2012). An overview of the Internet of Things for people with disabilities. *Journal of Network and Computer Applications, 35*(2), 584–596.

20. Melillo, P., Scala, P., Crispino, F., & Pecchia, L. (2014). Cloud-based remote processing and data-mining platform for automatic risk assessment in hypertensive patients. In *International Workshop on Ambient Assisted Living* (pp. 155–162). Springer International Publishing.

21. Pellicer, S., Santa, G., Bleda, A., Maestre, R., Jara, A., & Gomez Skarmeta, A. (2013). A Global Perspective of Smart Cities: A Survey. *Innovative Mobile and Internet Services in Ubiquitous Computing (IMIS), 2013 Seventh International Conference on,* 439–444.

22. World Bank (2016). World Development Report 2016: Digital Dividends. Washington, DC: World Bank. doi:10.1596/978-1-4648-0671-1. License: Creative Commons Attribution CC BY 3.0 IGO. Available online, http://documents.worldbank.org/curated/en/896971468194972881/310436360_201602630200228/additional/102725-PUBReplacement-PUBLIC.pdf [Accessed 10-Oct-2016].

23. BI Intelligence (2017) The smart cities report: Driving factors of development, top use cases, and market challenges for smart cities around the world. Available online, https://www.businessinsider.com.au/the-smart-cities-report-driving-factors-of-development-top-use-cases-andmarket- challenges-for-smart-cities-around-the-world-2016-10?r=US&IR=T [Accessed 10-Apr-2017].

24. Coetzee, L., & Olivrin, G. (2012). *Inclusion through the Internet of Things.* INTECH Open Access Publisher.

25. IControl Networks. (2015) State of the Smart Home Report. Available online: https://www.icontrol.com/wpcontent/uploads/2015/06/Smart_Home_Report_2015.pdf

26. Taylor, A., Harper, S., Swan, R., Izadi, L., Sellen, S., & Perry, A. (2007). Homes that make us smart. *Personal and Ubiquitous Computing, 11*(5), 383–393.

27. Brandt, S., Samuelsson, K., Tytri, O., & Salminen, A. (2011). Activity and participation, quality of life and user satisfaction outcomes of environmental control systems and smart home technology: A systematic review. *Disability & Rehabilitation: Assistive Technology, 2011, Vol.6(3), P.189–206, 6*(3), 189–206.

28. Li, R., Lu, B., & McDonald-Maier, K. D. (2015). Cognitive assisted living ambient system: a survey. *Digital Communications and Networks, 1*(4), 229–252.

29. ALM, N., & ARNOTT, J. (2015). Smart Houses and Uncomfortable Homes. *Studies in health technology and informatics, 217,* 146.

30. Sik-Lányi, C. (2015). Barriers and Facilitators to Uptake of Assistive Technologies: Summary of a Literature Exploration. *Assistive Technology: Building Bridges, 217,* 350.

31. Al-Fuqaha, A., Guizani, M., Mohammadi, M., Aledhari, M., & Ayyash, M. (2015). Internet of things: A survey on enabling technologies, protocols, and applications. *IEEE Communications Surveys & Tutorials, 17*(4), 2347–2376.

32. Gubbi, J., Buyya, R., Marusic, S., & Palaniswami, M. (2013). Internet of Things (IoT): A vision, architectural elements, and future directions. Future generation computer systems, 29(7), 1645–1660.

33. Sheng, Z., Yang, S., Yu, Y., Vasilakos, A., Mccann, J., & Leung, K. (2013). A survey on the ietf protocol suite for the internet of things: Standards, challenges, and opportunities. IEEE Wireless Communications, 20(6), 91–98.

34. Whitmore, A., Agarwal, A., & Da Xu, L. (2015). The Internet of Things—A survey of topics and trends. Information Systems Frontiers, 17(2), 261–274.

35. United Nations Department of Economic and Social Affairs, Population Division (2015). World Population Ageing 2015 (ST/ESA/SER.A/390). Available online, http://www.un.org/en/development/desa/population/publications/pdf/ageing/WPA2015_Report.pdf [Accessed 10-Oct-2016].

36. Meunier, F., Wood, A., Weiss, K., Huberty, K., Flannery, S., Moore, J., & Lu, B. (2014). *The Internet of Things Is Now: Connecting the Real Economy*. Technical Report.

37. Schindler, H. R., Schmalbein, N., Steltenkamp, V., Cave, J., Wens, B., & Anhalt, A. (2012). SMART TRASH: Study on RFID tags and the recycling industry.

38. Islam, S. R., Kwak, D., Kabir, M. H., Hossain, M., & Kwak, K. S. (2015). The internet of things for health care: a comprehensive survey. *IEEE Access, 3,* 678–708.

39. López, P., Fernández, D., Jara, A. J., & Skarmeta, A. F. (2013, March). Survey of internet of things technologies for clinical environments. In *Advanced Information Networking and Applications Workshops (WAINA), 2013 27th International Conference on* (pp. 1349–1354).

40. Yuehong, Y. I. N., Zeng, Y., Chen, X., & Fan, Y. (2016). The internet of things in healthcare: an overview. *Journal of Industrial Information Integration, 1,* 3–13.

41. Agu, E., Pedersen, P., Strong, D., Tulu, B., He, Q., Wang, L., & Li, Y. (2013). The smartphone as a medical device: Assessing enablers, benefits and challenges. In *Sensor, Mesh and Ad Hoc Communications and Networks (SECON), 2013 10th Annual IEEE Communications Society Conference on* (pp. 76–80).

42. Mosa, A., Yoo, I., & Sheets, L. (2012). A Systematic Review of Healthcare Applications for Smartphones. *BMC Medical Informatics and Decision Making, 12,* 67.

43. Abe, Y., Toya, M., & Inoue, M. (2013). Early detection system considering types of dementia by behavior sensing. In *Consumer Electronics (GCCE), 2013 IEEE 2nd Global Conference on* (pp. 348–349).

44. Ishii, H., Kimino, K., Aljehani, M., Ohe, N., & Inoue, M. (2016). An Early Detection System for Dementia Using the M2 M/IoT Platform. *Procedia Computer Science, 96,* 1332–1340.

45. Lawton, C. A. (1996). Strategies for indoor wayfinding: The role of orientation. *Journal of environmental psychology, 16*(2), 137–145.

46. BIANCHI, F., MASCIADRI, A., & SALICE, F. (2015). ODINS: On-Demand Indoor Navigation System RFID Based. *Studies in health technology and informatics, 217,* 341.

47. Hersh, M. A., & Johnson, M. A. (2008). Mobility: an overview. *Assistive technology for visually impaired and blind people,* 167–208. Springer London.

48. Mano, L. Y., Faiçal, B. S., Nakamura, L. H., Gomes, P. H., Libralon, G. L., Meneguete, R. I., & Ueyama, J. (2016). Exploiting IoT technologies for enhancing Health Smart Homes through patient identification and emotion recognition. *Computer Communications, 89,* 178–190.

49. Kao, H. L. C., Holz, C., Roseway, A., Calvo, A., & Schmandt, C. (2016, September). DuoSkin: rapidly prototyping on-skin user interfaces using skin-friendly materials. In *Proceedings of the 2016 ACM International Symposium on Wearable Computers* (pp. 16–23).

Chapter 5
Assistive IoT: Deployment Scenarios and Challenges

The impacts of physical and social settings on how empowered individuals may feel about achieving their potentials are well known. For people living with some sensory or cognitive impairment, including many older adults, the disabling or empowering effects of these settings are especially important. An inclusive environment that nurtures participation is based on considering the needs of an individual alongside the affordability and social constraints, employing available technology and knowledge in efficient manners. The IoT, with its pervasive nature and capabilities to provide and efficiently use intelligence in various forms and intensities, can be particularly of value in building inclusive and enabling smart environments. These environments can adapt to the needs and desires of people, track individuals to sense their intentions and objectives, inform the individual or others, and act to adjust some of the environmental aspects for the benefit of the persons [1].

Obviously, even if such environments could be fully realized, stigmas and biases still need to be overcome. New technologies, in general, and ICT revolutions, in particular, are embracing concepts of universal design that can be helpful for user adoptions and avoiding misconceptions surrounding age, impairments, and being different. However, the swift succession of ICT devices and services can have the danger of creating AT that is too expensive for most of their potential beneficiaries, suppressing the design for all strategies, or limiting their accessibility [2]. Besides these concerns, the familiarity of the users with the new technologies and steep learning curves can be hindrances to their usage. Concerns about losing privacy or reduced contact with familiar caregivers can also form obstacles for the use of the IoT-based smart environment by older adults or those living with disability. Lack of awareness of what is available and how they work is also one of the main issues for widespread adoption of relevant technologies. This chapter presents some of the implementations of the IoT and smart environments for aged care and empowering people living with dementia or some sensory impairment. It also discusses some of the challenges relevant to their deployment and usage.

© Springer International Publishing AG 2017
S. Shahrestani, *Internet of Things and Smart Environments*,
DOI 10.1007/978-3-319-60164-9_5

5.1 Assistive IoT: What Is Available, Feasible, and Missing

The majority of the proposed assistive IoT-based systems providing services for the elderly or individuals with sensory or cognitive impairment use body sensors and specific devices, while smart homes are clearly the new trend. An excellent survey and summarized comparison of the IoT-related systems is reported in [3]. A proprietary system, proposed by Intel, for example, is based on monitoring movements of the elderly to keep them safe [4]. However, it appears to lack any other sensing or functional capabilities. The smart experimental spaces that have been designed for evaluating how the sensing of assistive homes can be used to enhance the lives of the elderly are also proposed and built [5, 6]. Issues with sensing, communications, and proper interfacing seem to be at the core of the IoT-based assistive environments.

Many of the proposed solutions are based on implementing IP-based Wireless Sensor Networks (WSN) communicating over short ranges using WPANs, 6LoWPANs WBANs, WLANs and long-range systems, including broadband and cellular technologies. Furthermore, UWB, BLE, NFC, and RFID find widespread uses in application developments [7]. The IoT provides a suitable platform for AAL. Obviously, integration with wearable devices or other AI-based techniques can significantly enhance these platforms. Some samples of the assistive IoT-based systems are presented in this section.

IoT-based systems combine the pervasiveness of ubiquitous computing with the powers of the ambient intelligence to provide many features that can promote inclusiveness in smart environments. In particular, the following three characteristics are relevant to aged care and fostering a higher quality of life for people living with some sensory or cognitive impairment [1]. Through adjusting sensory information and proper interpretations of the environment, IoT-based environments make augmented reality possible. Such an augmentation can be highly useful for gaining insight into a physical environment and its virtualization to achieve advantages of the amalgamation of the cyber-physical worlds. The second feature relates to the distributed nature of many environmental things in providing more reliable sensing, actuation, and assistive services. Finally, the noninvasive characteristic of these environments can let people live their daily lives without noticing their assistive functionalities.

As these concepts are not yet mature enough, they undergo rapid changes with quick development successions. To improve their acceptance and usability, different stakeholders need to be included in all stages of their developments and implementations. People with disability, seniors, caregivers, and close family members can provide valuable insights into the requirements and interfaces for employment in these environments [8]. Taking a universal design approach, allowing people with diverse abilities to benefit from the devices and technologies that make up the environment, can also increase their adoption. The prospects of the acceptance of the IoT and smart environments depend on the availability of diverse types of sensors, actuators, smart devices, and appliances that use suitable interfaces, enabling them to directly communicate and interact with each other as well as with humans.

Identifying the needs of people living with some cognitive or physical impairment and satisfying those needs, is important for anyone, but can be particularly critical for those living alone. In some cases, this can simply relate to automatically raising the alarm in cases of accidents or emergencies. With an aging population, the decline of support from offspring and families, and high costs of formal care, smart homes and enabling environments can address such needs and have the capacity to go well beyond them.

Many studies have been carried out to identify the needs of older adults and those living with a sensory or cognitive impairment, and the suitable devices and solutions that can meet those needs. One example is the I-stay@home that incorporates an extensive European study that ran from 2012 to 2015, with the aim of identifying cost-effective solutions that may assist older people in living independently (http://istayathome.isah.aareonit.fr/index.php/Main_Page). The study installed ICT solutions in 180 homes of nine housing associations across Europe to evaluate products from more than 100 suppliers. The catalog of evaluated devices is available online (http://www.i-stay-home.eu/). Table 5.1 presents brief descriptions of some of those devices. Overall, the tenants and their caregivers were very pleased with the technology, with evaluations indicating that it can improve the quality of life, particularly for those living alone. A finding of this project has been the identification of significant market opportunities for large target groups. It found that many tenants have been able to implement ICT solutions without much difficulty, while significantly benefiting from them. For instance, the touchscreen interface of a tablet made it easy to use for most tenants. The project reports that gaining benefits from these solutions had little to do with age, gender, or physical abilities. The participants that did not have children or those who had above-average levels of education appeared to have benefitted the most. In the majority of the cases, the participations were the result of anticipation of the elderly staying longer in their own homes safely and comfortably.

The integration of some of these devices with AI and provisions of ambient intelligence in places of residence, hospitals, or other buildings can shape smart environments that are responsive to the needs of their users. To improve the quality of life for their users, they generally use the body and environmental sensors. Cameras can also be used to collect data non-intrusively. The data analyzed locally or remotely can generate information, alerts, or actions to be taken. The number of AAL-related developments and other smart home projects with assisted living goals is rapidly increasing.

Some of these projects have operated for some time, now. For instance, the EasyLiving project at Microsoft Research established in 1998 aimed to "…develop a prototype architecture and technologies for building intelligent environments that facilitate the unencumbered interaction of people with other people, with computers, and with devices..." (https://www.microsoft.com/en-us/research/publication/the-new-easyliving-project-at-microsoft-research/). Also, iDorm and iSpace are early research projects, running from 1999, on creating intelligent environments that employ fuzzy logic to learn from the behavior of the users, so that the spaces adapt to their needs (http://cswww.essex.ac.uk/research/iieg/idorm2/index.htm).

Table 5.1 Some of the smart devices evaluated to assist independent living of the older adults (http://www.i-stay-home.eu/)

VIVAGO [www.vivago.fr] wearable watch
• Manual and automatic alarms (automatic emergency call in case of loss of consciousness, hypothermia, or a fall)
• Can monitor sleep and activity of the user
• Detects if it not worn and can raise alarm
SOPHITAL [www.sophia.com.de] network
• Various functions for support at home
• Can automatically switch lights and outlets on or off
• Can monitor health or activity
• Can send alarm via e-mail or phone
iRobot Roomba [www.irobot.com] automatic, robotic vacuum cleaner
ComfiCare [www.comficare.nl] modular system
• Controlled through an app on a tablet
• Video communication
• Sharing information with caretakers, family, and relatives
• Digital control of lights, curtains, sunscreen, windows, temperature in various parts, and the like
• Control of the front door using the app
• Burglar and fire alarms that can be sent to others
TabTime Med-E-Lert [www.tabtime.com] automatic pill dispenser and alarm
GEOCARE [www.geo-care.de] mobile person locating system
• Emergency call button
• Can also raise the alarm if the user leaves a defined zone
LightwaveRF [www.lightwaverf.com] wireless control of lighting, power, and heating
• Fairly straightforward retrofitting of existing light switches, power points, heating and cooling controls with their wireless versions
• Controlled through an iOS or Android app
SALVEO [www.pervaya.com/en] movement detector and analyzer
• Array of motion sensors, not required to be worn
• Fall
• Can raise the alarm or contact a designated person
• Data can be used to analyze other risks (e.g., not rising from bed, irregular sleeping patterns, or abnormal toilet use)

House_n has been a long running multi-sponsored collaborative project at MIT, aimed at making homes that can benefit from adoption of digital technologies (http://web.mit.edu/cron/group/house_n/index.html).

The Gator Tech Smart House (GTSH) is an environment created by the Mobile and Pervasive Computing Laboratory of the University of Florida for experimentations with smart home technologies (https://www.cise.ufl.edu/~helal/gatortech/index2.html). The GTSH is an assistive environment focusing on the independence of the elderly population. One of its main advantages is the use of Atlas that facilitates the automatic integration of things, allowing for the development of smart environments without requiring engineering or system integration expertise [9, 10].

The Aware Home Research Initiative (AHRI) at Georgia Institute of Technology aims to "…develop requisite technologies to create a home environment that can both perceive and assist its occupants…" (http://www.cc.gatech.edu/fce/house/house.html). The technologies used in AHRI aim to promote interactions, cognition and context awareness, and management of emergencies and crises. It is a two-story residential building, similar to an average living house with the usual spaces and amenities. The experiments of short durations with actual residents living in an authentic home environment have indicated satisfactory results with regards to the use of the developed prototypes [3].

A rather substantive project funded by, US National Science Foundation (NSF) and run by Carnegie Mellon University and the University of Pittsburgh, is The Quality of Life Technology (QoLT) Centre. The main objective of this project is the development of smart devices and AT that enable independent living of older adults and people with disability. Some of the experiments run under the QoLT umbrella project, include prototyping personal assistive robots, cognitive and behavioral coaches, context awareness, driver AT, and efficient deployment of human-machine interactions [11].

There are also several systems that focus on supporting the informal carers. For example, by monitoring of daily care activities, the iCarer aims to provide an individualized and adaptive platform to supports informal caregivers of the elderly (http://www.aal-europe.eu/projects/icarer/).

ALADDIN is a platform that is more specifically designed to assist people with dementia, as well as their caregivers. Through regular monitoring, it can detect the early signs of dementia and manage the related potential emergencies with the goal of prolonging the period of safe in-home care [12]. A similar approach for the early detection of dementia in older adults living at home, based on using sensor technology, is used in BEDMOND. This system can convey the data to health professionals and continues to monitor the individual to evaluate the effectiveness of their treatments [13].

Another project with a similar title, ALADIN, is an ambient assisted living project under the European Framework Programme [14]. It is one of more than 100 AAL projects under this overarching EU Programme. ALADIN is based on the consideration that people's inner clock tends to deteriorate with aging, affecting their sleep-wake cycle regulation. It aims to use adaptive lighting to overcome the adverse effects that go with such deteriorations. ALADIN uses sensors to adaptively capture and analyze various effects of lighting on individuals and enable them to tailor the environment to meet their individualized needs. Several other works that focus on automatic sleep monitoring have also been carried out, for instance, see [15, 16].

PERSONA is another notable AAL project funded by the EU aiming to develop solutions for independent living of the older adults [17]. To achieve its outcomes, the project is divided into four areas, each relating to the category of AAL supportive services, namely, social inclusion, daily life activities, confidence and safety, and mobility [11].

ROSETTA is yet another project in the EU Programme, aiming to assist people with mild cognitive impairment and dementia. It is a modular system that can be of assistance to these individuals in carrying out their daily life functions, while it uses ambient sensors and video surveillance to monitor their daily behavior to identify emergency circumstances automatically [18]. Among the many functions of ROSETTA, surveyed users and informal carers, have referred to help in cases of emergencies, navigation support, and the calendar functions to be its most useful aspects.

Another project with similar goals is Geriatric Ambient Intelligence (GerAmi), which is a multi-agent based system for improved management of geriatric residences [19]. It incorporates mobile, Wi-Fi, and RFID technologies, along with case-based planning to reduce the burden of care through improved communication between the individuals and health professionals, as well as better task management. A prototype of the system has been tested at a facility that cares for AD patients.

IoT-based systems have also been widely anticipated to be capable of providing assistance to those with sensory or cognitive impairments. As discussed in Sect. 2.3, independent travel or moving around requires indoor and outdoor navigation capabilities. The blind and visually impaired people may rely on some type of AT to supplement their navigational abilities. There are many interesting, yet mostly in progress, IoT-based advances in this area. Many of these projects utilize some form of environmental sensors, tagging, localization, along with the sensors, computing power, and the interfaces of smartphones.

There are some clear advantages in using technologies for dual purposes or even incorporated in off-label solutions. For example, several projects are based on using the variations of signal strengths from several WLAN Access Points (AP) for localization of a mobile device that is connected to a person or an object. Such an approach to localization is, of course, a secondary usage of the existing WLAN that is already installed to provide wireless networking capacities in many buildings and environments. To work properly for localization purposes, rather accurate knowledge about the parameters like the locations and broadcasting powers of the APs are required. However, some of the data for these parameters are not usually measurable with the required accuracy, or they may dynamically change significantly. There are also problems associated with multipath propagation and secondary signals reflected from the walls, furniture, humans, and other objects in the environment. Some works account for these interferences, through calibrating the localization processes that use artificial neural networks [20]. While newer and more powerful approaches to deep learning may overcome such issues, with their deployments the core advantage of not using additional devices or complex computing approaches is already gone.

The combination of information from various sensors can provide more appropriate solutions. For example, smartphones and wireless sensors can be put to work together for building an indoor navigation system [21]. This system, same as others that use triangulation and the signal strengths of the surrounding base stations for positioning, can be affected by multipath propagations that render them imprecise and unreliable. However, with the advances in wireless and cellular technologies, they can be further developed in the context of the IoT. As UWB has relatively narrow RF pulses, its direct path and reflected signals can be distinguished from each other. As such, they have been used to obtain the required positioning information rather accurately [22]. This system requires UWB tags, smartphones and headphones providing the user interfaces, and the installation of sensors in the environment. The cost and specialized requirements can be the limiting factors for many situations.

The advanced sensors of the smartphones can also be utilized in providing meaningful interactions with the environment for individuals with different abilities. As mentioned in Sect. 4.1, a typical modern smartphone has 20 or more sensors. These may include GPS-based systems that can be useful for outdoor navigations, but in general, may lack the required precision and reliability for use by the blind or partially sighted individuals. They also include magnetic sensors and magnetometers that can act as a compass, which can be of value for indoor localization [23]. Gyroscopes and accelerator sensors that can detect the rotation and movement of the mobile device are almost invariably included in the smartphones. There are also cameras, proximity sensors, IR detectors, NFC and Bluetooth connections and beacons, as well as cellular trilateration for position capabilities, pressure meters, and possibly several others. While the number and precision of the sensors are different among the manufacturers and depending on the prices of the devices, they can only improve in the future. Several of these sensors are highly relevant to navigation and positioning of the device, and hence, a blind or visually impaired user who is carrying them.

The attractiveness of solutions that are based on using the capabilities of the mobile phones and the smartphones partly stems from their widespread use. World Bank has estimated that there were 2.33 billion 3G and a further 757 million 4G subscribers by the end of June 2015 [24]. As shown in Fig. 5.1, ITU estimates that

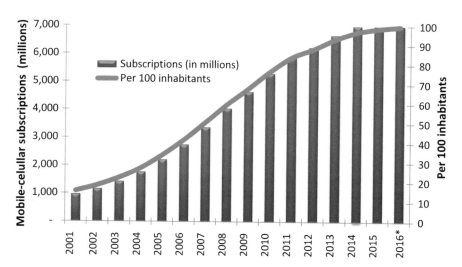

Fig. 5.1 Global mobile-cellular subscriptions, total and per 100 inhabitants, 2001–2016. Data Source: [25]

the total number of cellular subscriptions to be close to global population [25]. Around two billion of the subscribers are estimated to be using smartphones. The capabilities of mobile phones, particularly 3G and beyond, which can integrate their sensing and computing power with broadband connectivity to the Internet, provide many opportunities with positive impacts. They can also provide crowdsourced information that can be a good source of practical perception of the environment for an individual with some sensory or cognitive impairment. A number of recent works that highlight the usefulness of mobile phones, smartphones, and approaches that are based on their utilization are reviewed in [26]. That work has also pointed to several shortcomings of these solutions.

A promising line of works for navigation by the blind and partially sighted people is based on the use of Bluetooth beacons and audible instructions delivered through an interface device, such as bone conduction earphones or smartphones. A trial of such a system, "with 16 Bluetooth beacons providing pin-point accurate indoor location mapping" for unassisted mobility in the London Underground has been reported [27]. The system is based on a mobile app, Wayfinder, that promises to be an open platform promoting future developments [28]. Bluetooth beacons are employed to position the user by the app that provides audible directions to them through a bone conduction earphone.

Microsoft, Guide Dogs, UK and several other organizations have also embarked on expanding similar concepts to respond to the challenges that people with sight loss face while navigating the cities [29]. The developed technology includes an easy to use app, SoundScape, that has features allowing the user to start a journey in Orientate mode, and to explore or use Look Ahead mode, as well as to Find the Way mode, which lets them stop at any point and check that they are heading in the right direction using directional audio. A screenshot of the Orientate mode is shown in Fig. 5.2 (https://news.microsoft.com/en-gb/2015/11/25/cities-unlocked-a-voyage-of-discovery/#zFU7yRST1TEmvubV.99). The system, which is based on the inclusive design principles, is to be used as an AT alongside guide dogs and long canes, Fig. 5.3 (https://news.microsoft.com/stories/independence-day/).

There are several experiments and proposals for using intelligent acoustics to support applications in smart cities [30]. These may also be of potential benefit to people with hearing impairment. As depicted in Fig. 2.7, hearing loss affects 5% of the world population at disabling levels. The impairments and their effects usually get more prominent with age. For many cases, hearing aids, and in some situations, cochlear implants can provide some assistance to the affected people. Utilizing digital signal processing, powerful miniaturized CPU, and capability to handle some wireless technologies, these devices have progressed significantly over the past decade, or so. The communication technologies make them capable of working with many smartphones, TVs, and similar devices [31]. However, as these technologies have not been designed to work with these applications effortlessly, their adoption is not as high as originally anticipated [32]. Nevertheless, the development of a hearing aid profile for Bluetooth can improve the performance and deployment of these devices significantly [33]. That will also make these devices more compatible for use in an IoT-based smart environment.

Fig. 5.2 Orientate mode,
Cities Unlocked. Used
with permission from
Microsoft. Source: https://
news.microsoft.com/
en-gb/2015/11/25/
cities-unlocked-a-voyage-
of-discovery/

Fig. 5.3 A user of SoundScrape, whose experience has made him hopeful of more thoughtfully designed technology. Used with permission from Microsoft. Source: https://news.microsoft.com/stories/independence-day/

Recent releases of Bluetooth Low Energy (BLE) are meant to make hearing aids more accessible and provide more rewarding experiences for their users. BLE is to deliver easy to use and convenient connectivity between hearing aids and devices like smartphones, TVs, tablets, smart watches, and others that may be used in classrooms, public transport, or airplanes to provide the user with high-quality hearing experience. This integration can make the hearing aid devices much more affordable. The devices can be designed in the form of other smart wearables that can also be of benefit to the mainstream population. This market can then facilitate even more rapid developments.

For taking full advantage of the BLE in hearing aids, some obstacles need to be overcome. Perhaps, the most important one relates to the power requirements of the BLE. The battery size of typical hearing aid devices can prohibit the long-term use of the BLE chips. A solution can be to do most of the processing in another device, such as a smartphone, and stream the sound to the hearing aid. This solution requires the streaming to be supported by the BLE. Considering these points and noting that the only other existing wireless reception standard for hearing aids is the telecoil, which dates back to the 1950s, the Bluetooth Special Interest Group (SIG) and European Hearing Instrument Manufacturers Association have joined forces to accommodate BLE in "hearing aid use case requirements, including compatibility, audio quality, security, and power…" (https://www.bluetooth.com/news/pressreleases/2014/03/12/bluetooth-sig-and-ehima-partner-to-advance-hearing-instrument-technology-to--improve-the-lives-of-the-hearing-impaired).

Several major manufacturers have already introduced smart hearing aids to the market. For example, Siemens Primax smart hearing aids can be personalized with the touchControl App that makes a smartphone act as the hearing aid remote controller. It can be paired with easyTek streaming devices and app to connect the hearing aid to any Bluetooth-enabled device to act as advanced wireless headsets (http://www.businesswire.com/news/home/20160502005689/en/Signia-Introduces-primax-Advanced-Generation-Smart-Hearing). ReSound also manufactures hearing aids that can connect to other smart devices to stream all sounds and work like wireless stereo headphones (http://www.resound.com/en-AU/about-resound/smart-hearing). Starkey has marketed Halo 2 Made for iPhone hearing aids that use their TruLink 2.4 GHz wireless technology to enable connectivity with a variety of devices, to provide for wireless streaming of phone calls and music. The users can control the device to fit their individual needs in any environment (http://www.starkey.com.au/hearing-aids/technologies/halo-2-made-for-iphone-hearing-aids).

Connectivity of the hearing aids can also offer other significant advantages. For instance, processing of data and sounds can be carried with much greater computing power than what is available on a small device, like a behind the ear hearing aid. A major shortcoming of most hearing aids, relates to the poor comprehension of speech by a user in noisy environments, like in a restaurant or on a bus or a train. If some AT can overcome this problem, the market size can get tremendously bigger, as such devices may also assist people without any apparent hearing loss to better follow conversations in noisy everyday environments. One successful example of such an approach has shown to increase the ability to comprehend spoken words that were obscured by noise, by hearing impaired individuals, from scores near zero to values above 70% (http://asa.scitation.org/doi/10.1121/1.4820893). This work has used a classification algorithm that after some training, significantly improves the separation of speech from noise. Using deep machine-learning concepts, neural networks are trained to distinguish speech from noise using 85 attributes, including the sound frequencies and intensities. The reported results indicate that for individuals with some hearing impairment, comprehension of words scrambled by babble increased from the 29% baseline to 84% with the processing. For people who are not considered to be hearing impaired, the understanding of 42% of the words masked by babble improved to comprehending 78% of them with the use of this program [34].

The elderly and individuals with sensory or cognitive impairment can also benefit from the advances in robotics. Many AAL works are based on these technologies. Robots can be of value in assistance with mobility, daily living activities, cognitive abilities, and communication capacities [11]. This area is not that new, but many advances and interesting works are forthcoming [35]. For instance, Pearl is a robotic assistance for reminding the elderly of routine daily activities and helping them with safe navigations, using speech, visual display, and other easy to use interfaces [36]. RoboCare is an integrated distributed multi-agent system that provides domestic environment assistance for the older people living at home (http://robo-care.istc.cnr.it/). Many other assistive robots for elderly care are available. A review of the effectiveness of several of them has indicated positive outcomes, although backed by limited scientific evidence [37, 38].

The IoT and smart environments can also provide the basis for integration of wearable sensors, proximity sensors, and the smartphones and its gadgets. In this context, the added ambient intelligence enhances any user's interaction with the cyber-physical domains. The added interaction can, for example, assist visually impaired or blind individuals, with mobility and navigation, alerts and warnings, or identifying people or event of interest nearby [39]. These setups face the issues of dealing with the resources needed for handling a large number of sensors, or fast image processing and other analytics. Cloud-based approaches and upcoming techniques discussed in the previous section may be able to address some of these issues.

Various wearable devices have been mentioned in the previous chapter and other earlier parts. Their rapidly growing market is filled with many innovative solutions. Their location in proximity to the body and skin makes them suitable for taking some of the physiological readings, like the blood pressure, heart rate, or temperature. They can also track physical activities, sleep patterns, UV exposures, and respiratory rates to be analyzed for calories burnt or mood changes, for example. They can also be easily used for avoiding safety risks that some of the older adults or those living with dementia, face. They can be used as fall detectors, capable of raising alarms. They can provide non-obtrusive localization of their wearers, reducing the anxiety of the carers of people who may wander or get disoriented. Some of the low-cost wireless sensor networks that are easily wearable and can achieve these objectives are reported in [40, 41].

The array of wearable sensor network applications and use cases is rapidly expanding. This work can only list some samples of those cases. A wearable wireless electrocardiogram, ECG sensor, aiming for comfort and improved power efficiency, so that it can continuously monitor cardiac patients is proposed in [42]. A wearable ultrasonic sensor network that can be useful and cost-effective for routine clinical assessment of human arm motion is reported in [43]. Another wearable body area network uses ZigBee and Android-based smartphones to recognize various physical status and activities of humans [44]. Wearable systems, based on inertia measurement units positioned on the users' legs that use BLE, have also been used as non-intrusive mechanisms for early diagnosis of Alzheimer and Parkinson disease [45]. Continuous measurement of the Electrocardiogram and related physical activities that are carried out by wearable devices that use BLE to communicate them with a smartphone for analysis have been reported in [46].

The functioning of wearables and most other IoT-based technologies usually depend on data collection, communicating it with other devices, and carrying out some analysis. Industry players have also facilitated the ease of collection and usefulness of the data analytics. For instance, in 2014, Microsoft Health announced a combination of cloud-based predictive analytic services, its HealthVault, and measurements from the tracking devices to provide actionable insights that can be shared with medical professionals on a security-enhanced platform (https://www.microsoft.com/microsoft-health/en-au). In the same year, Apple announced HealthKit (https://developer.apple.com/healthkit/). It is meant to act as a platform that third-party fitness products and apps can use to "…become a valuable health data source and … use the shared data to bring more powerful health and fitness solutions...". Many other firms, supported by governmental policies, have made similar platforms that should contribute to the adoption of the wearables and the assistive IoT-based systems [47].

There are, however, as it will be discussed in the next section, many obstacles that make moving forward, challenging. As surfaced in the previous parts, security risk and privacy concerns probably top the list. The issues with interoperability of the huge number of things from different manufacturers and providers using different communication protocols and mechanisms is another concern that needs to be addressed. The unclear and varying market size predictions, and more specifically relevant to this work, the acceptability of the IoT-based AT solutions by the intended users, are yet the other main challenges.

5.2 Assistive IoT and Smart Environments: Challenges and Barriers

As pointed out in Chap. 2, many researchers believe that the focus on the technical aspects and functionality, at the expense of ignoring the human and social dimensions of AT, may have hindered the development of proper assistive devices and services. Many AT devices stay as prototypes. Very few of the promising modern devices are in actual use, not coming close to the uptake of the long cane or the guide dog. While the assistive IoT and smart environments may offer unique opportunities, they may share similar issues with other contemporary approaches, hindering their adoption and widespread employment. They may also face specific challenges. This section discusses some of the relevant matters, focusing on the specific challenges in moving forward.

With all the advantages that the solutions based on the IoT and smart environments offer, in general, they have not always enjoyed rapid adoption and growth.

For instance, the developments of smart cities are slowed down by budgetary constraints, bureaucratic processes, lack of clear policies or guidelines, and falling behind in priorities compared to the day-to-day running of a metropolis [48]. The slowdown is despite the enthusiasm of the public and political leaders in many parts of the world for embracing the smart infrastructures and improving the offered services. In some situations, the lack of understanding of the underlying issues that are mostly of technical or social nature, have resulted in politicizing the advances of the smart buildings or cities. In some ways, they have become controversial. Nevertheless, smart cities are seen as innovative and cost-effective means to revolutionizing the management of metropolises and related services. On the other hand, skeptics argue that the limited available resources must be invested in more fundamental and proven ways, rather than on extravagant technology and hype [49].

For the right or the wrong reasons, all new phenomena and technologies, particularly those affecting large parts of the societies are subject to such controversial views. IoT-based smart environments, AAL, and AT are no exceptions. As already shown, the IoT, wearable devices, and smart environments have resulted in the development of many new or enhanced products and services for assisted living. While the adoption of some of these devices and services have been high, not all of them have enjoyed that success. Even some of those that have seen strong market shares have not grown in a continuous and sustained fashion.

One of the issues, in many cases, is the lack of a holistic vision to guide the development and integration of such products and services with larger existing or evolving systems. Such integrations can facilitate the collection and processing of large volumes of data, capability to analyze complex situations, and ability to provide collaborative solutions [11]. As it stands, with the very fast pace of expansions and growth, many of the devices or services seem to be developed in isolation, with connectivity and integration being after-thoughts. Hence, same as with the Internet itself or to a lesser extent with cellular phones and networks, security and privacy, interoperability, and appropriate interfaces are among the major growth requirements, not properly considered and resolved in the original systems.

IBM sees the IoT as the largest single source of data globally (http://www.ibm.com/internet-of-things/). Such big data holds significant potential values that can be turned into innovations and operational efficiency through utilization of analytics. The combination of big data analytics and the IoT provides assistive living solutions that can be holistic in nature. However, the overwhelming amount of data can also disenfranchise the humans, who are part of this ecosystem. Generally speaking, it is hard for most people to see how to make the best use of the amount of data and information that is already available. Such difficulties may even include the selection of appropriate devices and services that meet their needs. The age-old information overload can be a major issue with the IoT, as in its core the advances of the IoT are based on its power to automate the generation, processing, and utilization of data. This management of everything by machines, appearing as the lack of direct human influences, can cause fundamental problems, for instance, related to mechanical and superficial personalization [50].

Problems that deal with the use of data can also arise from rapid growths in some areas, which have not been matched by others. In the case of the IoT and smart environments, the advances in sensor technologies and abilities for comprehensive monitoring and surveillance have not been complemented by growth in other computing resources and interface technologies. While the progress is rapid in all of these fronts, data availability still surpasses the processing and use capabilities. For instance, real-time localization and navigation systems that can assist the blind or visually impaired people to travel around, face two major related challenges. The first one is the allocation of computational resources that can process the large amounts of data coming from multiple sensors and cameras, fast enough in a synchronized manner [51]. The second issue relates to quick and real-time access to dynamic datasets through interfaces that are appropriate for the user.

From a technical perspective, compared to more conventional networking devices, the things and the IoT objects, and henceforth AAL, smart environments, and the AT that are based on them have several limitations. Most IoT devices are energy constrained, with small memory, limited processing power, and restrictive communication capabilities. The sheer number of IoT devices, mandates the use of IPv6, requiring integration of multi-technology networks and managing IPv4 and IPv6 coexistence issues [52]. Finding cost-effective solutions to all of these problems can be difficult.

Cost can obviously be a factor in the adoption of any technology. The expenses associated particularly with the more advanced AT, make them out of reach for many elderly and people living with dementia or disability. It is a well-known unfortunate fact that global hearing aid productions meet less than 10% of the need, and in developing countries, only 3% of the people who need them, can access them [53]. While this can especially distress people living in low and middle-income countries, individuals in developed countries are also proportionally affected. In the UK, for example, the main reason for not using the Internet, given by people with disability, is the costs involved, including those associated with acquiring the required assistive devices and technology, such as a screen reader [2].

Another aspect of the AT adoption relates to its suitability in meeting the users' needs and desires. A critical factor to consider is that some approaches "…may have technical merit, and may solve obvious problems, but still fail to address the complex interplay of issues at work and to take the most appropriate approach to dealing with these matters. Furthermore, it is important to acknowledge that there may not even be a 'right' problem to tackle. Flexibility cannot be overvalued…." [54]. In this sense, to develop devices and services that are actually used, a holistic approach that incorporates the mental state and wishes of the intended users, is required. Here again, the lack of a comprehensive approach can result in products that only serve a perceived or assumed need, rather than filling an actual gap. For instance, some may simply ignore the mentality of an elderly, who may feel that they have no place to go, even after subjecting themselves to the distresses associated with using an advanced AT that may assist them with mobility or travel [55]. As another example, consider the smart doorbell project that was mentioned in Sect. 3.3. It is meant to recognize individuals at the door and notify the user through their mobile devices. The adoption and employment of such systems, clearly depends on the confidence of the user of their error rate, as they do not want to miss anyone whom they like to let in [56].

Similar accounts also hold for some wearables. Many of the great forecasts for them have not eventuated. For instance, IDC reports have shown that by November 2016 the year-on-year shipments of smartwatches has declined by nearly 52% [57]. Microsoft has removed its Fitness Band from its online store, no longer providing the Band developer kits. Some analysts have contributed the decline in the wearables market to the lack of new or engaging capabilities. Technical issues like the battery power issues and dependence on other devices like smartphones have also been mentioned, with the keys to growth identified as providing actually needed functionalities and capability of direct connection to the cellular networks [58]. The case for smartwatches has its own specific considerations, too. They need to be kept small to be worn on the wrist without being too obtrusive. As such, the battery life, processing power, interface features, and its sensors will have limited capabilities compared to other smart devices, such as smartphones [59]. These limitations result in inferior functionalities that need to be overcome. Besides these, some of the functionalities of these devices have not actually led to desired outcomes. For example, a 24-month study has reported that adding wearable devices to standard behavioral interventions, has resulted in less weight loss [60]. While many of these devices can track and even promote physical activity, they may not be able to offer significant advantages regarding weight loss that is the final goal for many users.

Another relevant issue in taking a holistic view of AT developments, besides the functionally and being suitably designed to meet their purposes, is the stigma that may go with an AT. If the AT design is such that its employment can make an individual seem like they may not meet the implied expectations of others, it can add to the discrimination and inaccessibility barriers [61]. The discrimination can, in turn, result in rejection of that AT by its intended users. A holistic approach needs to consider technical aspects, experiences of various stakeholders, social acceptance, avoidance of stigma, and other human elements. Successful adoption of the new AT depends heavily on its consideration of this wider view and context. This dependency can, for example, be seen by the relatively small number of electronic travel aids that are used by the blind and visually impaired people, compared to the widespread use of the long cane and guide dogs. Some of the possible reasons for the lack of success of the newer AT can be related to their difficulty of use, their high costs, their inappropriate appearances, not meeting the users' needs, being awkward or heavy to carry, and not providing superior functionality compared to the more traditional AT [62].

Some people may feel intimidated by the newer technologies such as those used in the smart environments and the IoT-based AT. Most individuals are used to being in control of their devices, whereas IoT-based AT and environments tend to automate processes, ironically, to alleviate the pressures of operating devices and appliances for the humans [50]. A similar issue arises from the fast pace of the development of such advanced technologies. The rapid technological advances, make the products and services obsolete, as the new improved versions become available at fast rates. These rapid changes usually mean that the elderly or people with disability or dementia may miss out on obtaining the full benefits of these devices or services [2]. The costs, learning curves, or simply a lack of awareness can also mean that these people may not use the new technologies. Some may not even try to identify newer AT, such as those based on the IoT, for the older people, as they are viewed as being reluctant to engage with the newer technologies. The degree of validity of such views remains to be established with solid data. A 2012 study has, however, indicated that due to such perceived reluctance, merely just over 20% of people caring for a person with dementia were going to look for relevant information on the Internet [63].

Many people are wary of the invasion of their privacy by the IoT and connected devices and appliances. News about the smart TVs that could record conversations in the room and send them over the Internet to a covert CIA server have surfaced and caused privacy concerns for some people (http://www.cnbc.com/2017/03/09/if-you-have-a-smart-tv-take-a-closer-look-at-your-privacy-settings.html). If true, the problem for smart TVs, is not the lack of processing power. However, generally speaking, the IoT devices lack the processing power for being fully secured, while they have some processing capability and are connected to other things or the Internet. In both cases, the issue may relate to the technology that is designed in such a way that it may not alert the user that a device is collecting data and potentially sharing it with third parties. Even if it did though, the expectation that inexperienced users can remedy the issue is unrealistic. These ideas are part of the modern culture, with stories about the attack by the fridges [64].

The security and privacy challenges for the AT and AAL based on the IoT are serious. IoT devices can be used for cyber attacks, be targeted causing them to malfunction, or be hijacked to expose user data and information. They may have no set way to alert the user when a security problem arises [65]. Securing connected IoT devices is generally more challenging than securing smartphones or a laptop. In part, that is due to lack of standards for the IoT market that is yet to mature, with many small and new players that may not have enough expertise in dealing with security issues. Another complexity relates to the heterogeneity of the IoT devices. Many of these devices may be manufactured as low-cost and disposable products, making it hard or impossible to update their software or apply a patch to overcome their discovered vulnerabilities [47]. Furthermore, the majority of the IoT devices lack the basic security protections like the capability to use changeable strong passwords or encryption techniques [66].

It is not surprising that it has been indicated that one of the main criticisms about assistive IoT found in the literature has been the violation of privacy [50]. Finding the right balance between preservation of the privacy and giving users appropriate levels of control, while collecting enough data to provide the right assistance, can be difficult [67]. For example, consider provisions of mobility and wayfinding discussed in Sects. 3.3 and 4.4. These provisions require collecting information about the user location, as well as about their movements, intentions, and environments. The data needs to be shared with a higher level service to carry out the necessary analysis and provide the user with navigation information. Clearly, collecting and processing this amount of personal information can be considered by some users to be a breach of their privacy [1].

There are also a variety of potential security risks associated with the IoT. An FTC-run workshop and Staff Report has identified several of them [47]. They include unauthorized access to personal information, facilitating attacks on other systems, and creating risks to personal safety. It is noted that privacy risks may stem from the collection of personal data, positions, and health information. Unauthorized access to data from cameras or baby monitors, for instance, can raise potential physical safety concerns. Data from smart meters can be used to determine whether a place is occupied or not. The report gives an example of remote access into two different connected insulin pumps whose settings were changed, putting the personal safety of their users at risk. It makes the observation that perceived risks to privacy and security, whether realized or not, can easily damage the intended users' confidence, leading to lowering of the adoption and employment of the technologies.

Lack of awareness of available IoT-based AT is also a barrier to their adoption. This lack of knowledge is an issue with many newer AT and technological developments. In several studies, it has been found that participants were describing their needs for development of devices and services that were already commercially available. Survey results in a study in the UK, have indicated that even the primary and secondary dementia care physicians may not be aware of the latest development of the relevant AT [63]. This study has indicated that AT is not generally used by the those living with dementia or their carers. With a small sample size, it found it difficult to ascertain the barriers to AT use. However, it has reported that members of its focus group suggested AT development ideas that related to devices that already are available, indicating a severe lack of awareness by many stakeholders.

References

1. Coetzee, L., & Olivrin, G. (2012). *Inclusion through the Internet of Things*. INTECH Open Access Publisher.
2. World Health Organization. (2011). *World report on disability*. World Health Organization.
3. Mano, L. Y., Faiçal, B. S., Nakamura, L. H., Gomes, P. H., Libralon, G. L., Meneguete, R. I. & Ueyama, J. (2016). Exploiting IoT technologies for enhancing Health Smart Homes through patient identification and emotion recognition. *Computer Communications*, *89*, 178–190.
4. Ram, S. (2016). Internet-of-Things (IoT) Advances Home Healthcare for Seniors. Available online, http://www.embeddedintel.com/special_features.php?article=2721 [Accessed 10-Oct-2016].
5. Chen, C., Helal, S., de Deugd, S., Smith, A., & Chang, C. K. (2012). Toward a collaboration model for smart spaces. In Software Engineering for Sensor Network Applications (SESENA), 2012 Third International Workshop on (pp. 37–42).
6. Deen, M. J. (2015). Information and communications technologies for elderly ubiquitous healthcare in a smart home. Personal and Ubiquitous Computing, 19(3–4), 573–599.
7. Celentano, U., & Röning, J. (2015). Framework for dependable and pervasive eHealth services. In *Internet of Things (WF-IoT), 2015 IEEE 2nd World Forum on* (pp. 634–639).

8. Frauenberger, C. (2015). Disability and technology: A critical realist perspective. In *Proceedings of the 17th International ACM SIGACCESS Conference on Computers & Accessibility* (pp. 89–96).

9. Helal, S., Bose, R., Pickles, S., Elzabadani, H., King, J., & Kaddourah, Y. (2008). The Gator Tech Smart House: A Programmable Pervasive Space. *The Engineering Handbook of Smart Technology for Aging, Disability, and Independence*, 693–709.

10. Helal, S., Mann, W., El-Zabadani, H., King, J., Kaddoura, Y., & Jansen, E. (2005). The gator tech smart house: A programmable pervasive space. *Computer*, *38*(3), 50–60.

11. Li, R., Lu, B., & McDonald-Maier, K. D. (2015). Cognitive assisted living ambient system: a survey. *Digital Communications and Networks*, *1*(4), 229–252.

12. Federici, S., Tiberio, L., & Scherer, M. J. (2014). Ambient Assistive Technology for People with Dementia: An Answer to the Epidemiologic Transition. *New Research on Assistive Technologies: Uses and Limitations. New York, NY: Nova Publishers*, 1–30.

13. Fuxreiter, T., Mayer, C., Hanke, S., Gira, M., Sili, M., & Kropf, J. (2010). A modular platform for event recognition in smart homes. In *e-Health Networking Applications and Services (Healthcom), 2010 12th IEEE International Conference on* (pp. 1–6).

14. Maier, E., & Kempter, G. (2010). ALADIN-a Magic Lamp for the Elderly? *Handbook of Ambient Intelligence and Smart Environments*, 1201–1227.

15. Jiang, L., Tang, Z., Liu, Z., Chen, W., Kitamura, K. I., & Nemoto, T. (2012). Automatic sleep monitoring system for home healthcare. In *Biomedical and Health Informatics (BHI), 2012 IEEE-EMBS International Conference on* (pp. 894–897).

16. Zhu, X., Zhou, X., Chen, W., Kitamura, K. I., & Nemoto, T. (2014). Estimation of Sleep Quality of Residents in Nursing Homes Using an Internet-Based Automatic Monitoring System. In *Ubiquitous Intelligence and Computing, 2014 IEEE 11th Intl Conf on and IEEE 11th Intl Conf on and Autonomic and Trusted Computing, and IEEE 14th Intl Conf on Scalable Computing and Communications and Its Associated Workshops (UTC-ATC-ScalCom)* (pp. 659–665).

17. Fides-Valcro, Á., Freddi, M., Furfari, F., & Tazari, M. R. (2008). The PERSONA framework for supporting context-awareness in open distributed systems. *Ambient Intelligence*, 91–108.

18. Meiland, F., Hattink, B., Overmars-Marx, T., De Boer, M., Jedlitschka, A., Ebben, P., Dröes, R. (2014). Participation of end users in the design of assistive technology for people with mild to severe cognitive problems; the European Rosetta project. 26(5), 769–779.

19. Corchado, J. M., Bajo, J., & Abraham, A. (2008). GerAmi: Improving healthcare delivery in geriatric residences. *IEEE Intelligent Systems*, *23*(2).

20. Rodriguez, M. D., Favela, J., Martínez, E. A., & Muñoz, M. A. (2004). Location-aware access to hospital information and services. *IEEE Transactions on information technology in biomedicine*, *8*(4), 448–455.

21. Cecilio, J.; Duarte, K.; Furtado, P. (2015). BlindeDroid: An information tracking system for real-time guiding of blind people. Procedia Comput. Sci. 2015, 52, 113–120.

22. Martinez-Sala, A. S., Losilla, F., Sánchez-Aarnoutse, J. C., & García-Haro, J. (2015). Design, implementation and evaluation of an indoor navigation system for visually impaired people. *Sensors*, *15*(12), 32168–32187.

23. Haverinen, J., & Kemppainen, A. (2009). Global indoor self-localization based on the ambient magnetic field. *Robotics and Autonomous Systems*, *57*(10), 1028–1035.

24. Halewood, A. M. K., Sabino, M., Sudan, R., & Yadunath, D. (2015). Six digital technologies to watch. World Bank. Available online, http://documents.worldbank.org/curated/en/896971468 194972881/310436360_201602630200216/additional/102725-PUB-Replacement- PUBLIC. pdf [Accessed 10-Oct-2016].

25. International Telecommunication Union (2017). World Telecommunication/ICT Indicators database 2016. ITU. Available online, http://www.itu.int/en/ITU-D/Statistics/Pages/publications/wtid.aspx [Accessed 10-Feb-2017].

26. Csapó, Á., Wersényi, G., Nagy, H., & Stockman, T. (2015). A survey of assistive technologies and applications for blind users on mobile platforms: a review and foundation for research. *Journal on Multimodal User Interfaces, 9*(4), 275–286.

27. Pym, H. (2015). Technology helps visually impaired navigate the Tube. BBC News. Available online, http://www.bbc.com/news/health-31754365 [Accessed 10-Oct-2016].

28. Wayfinder. (2016). Wayfindr- Empowering vision impaired people. Available online, https://www.wayfindr.net/ [Accessed 10-Oct-2016].

29. Cities Unlocked. (2016). Cities Unlocked: Realising the potential of people and places. Available online, http://www.citiesunlocked.org.uk/ [Accessed: 10-Oct-2016].

30. Pham, C., & Cousin, P. (2013). Streaming the sound of smart cities: experimentations on the SmartSantander test-bed. In Green Computing and Communications (GreenCom), 2013 IEEE and Internet of Things (iThings/CPSCom), IEEE International Conference on and IEEE Cyber, Physical and Social Computing (pp. 611–618).

31. Swanepoel, D. W. (2017). Smartphone-based National Hearing Test Launched in South Africa. *The Hearing Journal, 70*(1), 14–16.

32. Garcia-Espinosa, E., Longoria-Gandara, O., Veloz-Guerrero, A., & Riva, G. G. (2015). Hearing aid devices for smart cities: A survey. In *Smart Cities Conference (ISC2), 2015 IEEE First International* (pp. 1–5).

33. Mecklenburger, J., & Groth, T. (2016). Wireless Technologies and Hearing Aid Connectivity. In Hearing Aids (pp. 131–149). Springer International Publishing.

34. Wang, D. (2017). Deep learning reinvents the hearing aid. *IEEE Spectrum, 54*(3), 32–37.

35. Mohammed, S., Moreno, J., Kong, K., & Amirat, Y. Eds.). (2015). (*Intelligent Assistive Robots: Recent Advances in Assistive Robotics for Everyday Activities* (Vol. 106). Springer.

36. Pollack, M. E., Brown, L., Colbry, D., Orosz, C., Peintner, B., Ramakrishnan, S., & Thrun, S. (2002). Pearl: A mobile robotic assistant for the elderly. In *AAAI workshop on automation as eldercare* (Vol. 2002, pp. 85–91).

37. Bemelmans, R., Gelderblom, G. J., Jonker, P., & De Witte, L. (2012). Socially assistive robots in elderly care: A systematic review into effects and effectiveness. *Journal of the American Medical Directors Association, 13*(2), 114–120.

38. Kachouie, R., Sedighadeli, S., Khosla, R., & Chu, M. T. (2014). Socially assistive robots in elderly care: a mixed-method systematic literature review. *International Journal of Human-Computer Interaction, 30*(5), 369–393.

39. Yelamarthi, K., Haas, D., Nielsen, D., & Mothersell, S. (2010). RFID and GPS integrated navigation system for the visually impaired. In Circuits and Systems (MWSCAS), 2010 53rd IEEE International Midwest Symposium on (pp. 1149–1152).

40. Caldara, M., Locatelli, P., Comotti, D., Galizzi, M., Re, V., Dellerma, N., & Pessione, M. (2015). Application of a wireless BSN for gait and balance assessment in the elderly. In *Wearable and Implantable Body Sensor Networks (BSN), 2015 IEEE 12th International Conference on* (pp. 1–6).

41. Pivato, P., Dalpez, S., Macii, D., & Petri, D. (2011). A wearable wireless sensor node for body fall detection. In *Measurements and Networking Proceedings (M&N), 2011 IEEE International Workshop on* (pp. 116–121).

42. Wong, D. L. T., & Lian, Y. (2012). A wearable wireless ECG sensor with real-time QRS detection for continuous cardiac monitoring. In *Biomedical Circuits and Systems Conference (BioCAS), 2012 IEEE* (pp. 112–115).

43. Qi, Y., Soh, C. B., Gunawan, E., & Low, K. S. (2014). A wearable wireless ultrasonic sensor network for human arm motion tracking. In *Engineering in Medicine and Biology Society (EMBC), 2014 36th Annual International Conference of the IEEE* (pp. 5960–5963).

44. He, Z., & Bai, X. (2014). A wearable wireless body area network for human activity recognition. In *Ubiquitous and Future Networks (ICUFN), 2014 Sixth International Conf on* (pp. 115–119).

45. Margiotta, N., Avitabile, G., & Coviello, G. (2016). A wearable wireless system for gait analysis for early diagnosis of Alzheimer and Parkinson disease. In *Electronic Devices, Systems and Applications (ICEDSA), 2016 5th International Conference on* (pp. 1–4).
46. Khan, J. A., Akbar, H. A., Pervaiz, U., & Hassan, O. (2016). A wearable wireless sensor for cardiac monitoring. In *Wearable and Implantable Body Sensor Networks (BSN), 2016 IEEE 13th International Conference on* (pp. 61–65).
47. Federal Trade Commission. (2015). Internet of Things: Privacy & security in a connected world. *Washington, DC: Federal Trade Commission.*
48. BI Intelligence (2017) The smart cities report: Driving factors of development, top use cases, and market challenges for smart cities around the world. Available online, https://www.businessinsider.com.au/the-smart-cities-report-driving-factors-of-development-top-use-cases-andmarket- challenges-for-smart-cities-around-the-world-2016-10?r=US&IR=T [Accessed: 10-Oct-2016].
49. World Bank (2016). World Development Report 2016: Digital Dividends. Washington, DC: World Bank. doi:10.1596/978-1-4648-0671-1. License: Creative Commons Attribution CC BY 3.0 IGO. Available online, http://documents.worldbank.org/curated/en/896971468194 972881/310436360_201602630200228/additional/102725-PUBReplacement-PUBLIC.pdf [Accessed 10-Oct-2016].
50. Nolin, J., & Olson, N. (2016). The Internet of Things and convenience. *Internet Research, 26*(2), 360–376.
51. Xiao, J., Joseph, S. L., Zhang, X., Li, B., Li, X., & Zhang, J. (2015). An assistive navigation framework for the visually impaired. *IEEE Transactions on Human-Machine Systems, 45*(5), 635–640.
52. Elkhodr, M., Shahrestani, S., & Cheung, H. (2016). Emerging Wireless Technologies in the Internet of Things: a Comparative Study. *arXiv preprint arXiv:1611.00861.*
53. World Health Organization. (2015). *WHO global disability action plan 2014–2021: Better health for all people with disability.* World Health Organization.
54. Mankoff, J., Hayes, G. R., & Kasnitz, D. (2010). Disability studies as a source of critical inquiry for the field of assistive technology. In *Proceedings of the 12th international ACM SIGACCESS conference on Computers and accessibility* (pp. 3–10).
55. Nihei, M., & Fujie, M. G. (2012). *Proposal for a New Development Methodology for Assistive Technology Based on a Psychological Model of Elderly People.* INTECH Open Access Publisher.
56. Hamann, L. M. A., Airaldi, L. L., Molinas, M. E. B., Rujana, M., Torre, J., & Gramajo, S. (2015, December). Smart doorbell: An ICT solution to enhance inclusion of disabled people. In *ITU Kaleidoscope: Trust in the Information Society (K-2015), 2015.* (pp. 1–7).
57. Kleinman, Z. (2016). Smartwatch sales show sharp decline, report finds. BBC News. Available online, http://www.bbc.com/news/technology-37762239 [Accessed 10-Oct-2016].
58. Kleinman, Z. (2017). Has wearable tech had its day? BBC News. Available online, http://www.bbc.com/news/technology-39101872 [Accessed 10-Oct-2016].
59. Rawassizadeh, R., Price, B. A., & Petre, M. (2015). Wearables: Has the age of smartwatches finally arrived? *Communications of the ACM, 58*(1), 45–47.
60. Jakicic, J. M., Davis, K. K., Rogers, R. J., King, W. C., Marcus, M. D., Helsel, D., & Belle, S. H. (2016). Effect of wearable technology combined with a lifestyle intervention on long-term weight loss: the IDEA randomized clinical trial. *Jama, 316*(11), 1161–1171. Available online, http://jamanetwork.com/journals/jama/fullarticle/2553448 [Accessed: 10-Oct-2016].
61. Shinohara, K., & Wobbrock, J. O. (2011). In the shadow of misperception: assistive technology use and social interactions. In *Proceedings of the SIGCHI Conference on Human Factors in Computing Systems* (pp. 705–714).
62. Hersh, M. A., & Johnson, M. A. (2008). Mobility: an overview. *Assistive technology for visually impaired and blind people*, 167–208. Springer London.

63. van den Heuvel, E., Jowitt, F., & McIntyre, A. (2012). Awareness, requirements and barriers to use of Assistive Technology designed to enable independence of people suffering from Dementia (ATD). *Technology and Disability, 24*(2), 139–148.

64. Grau, A. (2015). Can you trust your fridge? *IEEE Spectrum, 52*(3), 50–56.

65. Elkhodr, M., Shahrestani, S., & Cheung, H. (2016). The Internet of Things: new interoperability, management and security challenges, *International Journal of Network Security and Its Applications*, vol 8, no 2, pp 85–102.

66. Ahlmeyer, M., & Chircu, A. M. (2016). Securing the internet of things: A review. *Issues in Information Systems, 17*(4).

67. ALM, N., & ARNOTT, J. (2015). Smart Houses and Uncomfortable Homes. *Studies in health technology and informatics, 217*, 146.

Chapter 6
Assistive IoT and Smart Environments: Performance Requirements

Experts from various areas agree that eight domains somehow underline the quality of life for most human beings [1, 2]. These domains include emotional well-being, interpersonal relationships, financial and material security, personal development, physical health, self-determination, social inclusion, and rights. Traditional and advanced AT can improve the quality of life for people with diverse abilities in many of these domains [3]. The performance of AT in achieving such improvements depends on many requirements that need to be identified and met. These are covered in this chapter.

The disability models and their interactions have many implications for the technology. The usability and acceptance by the intended users may be even more important than the intended functionality. These factors, in turn, relate to the ease of use and the utilized interfaces, as well as to social acceptance and aesthetics [4]. The obvious example cited relevant to this point is the way people have accepted eyeglasses or contact lenses as accessories rather than medical aids.

The acceptance of the AT depends on some distinct aspects that several studies, some quite extensive, have identified [4, 5]. These include acceptance of the disability by the technology, social inclusion and if the device made the user stand out, and cultural values. They have shown that the users see the value of technology in providing access and capability to behave and appear like everyone else. They have, however, reported that AT lagged behind most other technologies in these respects. While acknowledging that no AT will fully replace sight, hearing, or other functions, access to information or facilities similar to everyone else, is valued.

People with disability may find it important to choose to disclose their disability or not. This choosing may be related to perceptions of others and the potential stigmas that go with them. As such, they may only use the AT that could help them achieve independence and show their abilities. The achievement-based usage of devices can be extended to the complexity of employing and interacting with the technology. Individuals may avoid the situations that may contribute to their capability misperceptions. Misperceptions about their abilities can be mitigated if affected people use the technology that everyone is using. Utilization of

S. Shahrestani, *Internet of Things and Smart Environments*,
DOI 10.1007/978-3-319-60164-9_6

similar-looking devices is particularly suitable and yet possible where, for instance, electronic AT can be designed to look the same as the conventional devices with comparable functionalities [5]. To meet such requirements, the IoT, smartphones, smart homes, buildings, cities, and other environments can be of great value. Accustomed devices, like smartphones, smartwatches, or smart TVs can be used to provide the required interfaces for interactions. The technology can be easily based on universal design concepts, inclusive of most individuals and groups of people. This chapter covers these topics.

6.1   Perception of the Surrounding Environment

Most people feel comfortable and even empowered when they are in their familiar settings. Environments that can sense the presence of an individual and respond to their needs can be really empowering. These are the basic concepts behind ambient intelligence and AAL applications. The ambient, as an information paradigm, can provide individualized and fine-tuned information related to the presence, tracking, and monitoring the person. These can be particularly empowering for the elderly, and for people living with disability or dementia. In more general terms, ambient intelligence can increase an individual's awareness levels. As discussed in Sect. 2.3, this can, for instance, provide the elderly with information about the events that can be of interest to them, inspiring social activities and inclusion [6]. Such intelligence can be used to identify and locate friend or events nearby.

The ambient intelligence can assist the blind or visually impaired persons with navigations and wayfinding. AAL and ambient intelligence have been used in many projects whose main aims are either navigation assistance or providing environmental awareness for the blind and visually impaired persons. Various interesting approaches have been suggested to achieve these goals. In almost all cases, wearable or mobile sensors, which complement the environmental sensors and tags, are used. The environmental awareness can relate to many features of interest, for example, used for discovering some social events of interest, changes in traffic conditions, or simply identifying a familiar face passing by.

By integrating cameras, proximity sensors, accelerometers, and gyroscopes with information from environmental sensors, social media, and crowdsourced from the IoT, a system that can alert a user of being in the vicinity of an event of interest has been developed [7]. The system, whose usability studies have been carried out in collaboration with visually impaired communities in New York City, provides a low-frequency sound alert with changing amplitude that corresponds to the distance to the event. It alerts the users to misadventures with a high amplitude and a high-frequency audio warning. For navigations, this system uses wearable sensors, SLAM, Internet access through Wi-Fi, and interactions with an intelligent transport network through a dedicated short-range communication technology. The mobile system consists of wearable sensors, feedback devices, and human-machine interfaces. It also contains a portable computing device, such as a tablet or a smartphone, with network connectivity to access resources, such as geographic information systems, transportation databases, and social media networks. Outdoor navigations rely on a GPS-based system and various software and applications. The indoor navigations depend on having access to an image of the floor plan, with obstacle-avoidance achieved by using a camera.

Other projects are based on similar architectures and concepts. They aim to enhance assistive navigation, crosswalk passing, context recognition, mapping, or floor plan digitization [7, 8]. A more specific AT project has developed a system that facilitates the independent use of public buses by visually impaired or blind people as well as by the elderly individuals [9]. The Bluetooth-based bus notification system attempts to remove the information barriers for these groups of commuters. The proposed system requires Bluetooth modules on all the buses and in all bus stops. So, unless the system is usable by many commuters, additional costs can render it economically unviable.

Navigation systems that are based on GPS are very popular for general use by sighted people, too. This large market has facilitated the growth and advancement of many related technologies and devices. Mobile phones and smartphones offer rather advanced navigation features, including maps and positioning systems. GPS technology uses satellite information, which is projected on a digital map to identify the position of the device and its relation to landmarks of interest. For most of the in-car navigation systems, the input is usually via touchscreen and the output is through a visual display and spoken information. Most of these features, particularly if enhanced with speech recognition capabilities, can be of great value to many people with different abilities.

For use by the blind or visually impaired people, the accuracy of the traditional GPS-based navigation may not be sufficient. There have been several approaches to increasing this precision. For example, Differentially corrected GPS (DGPS) can give accuracies in the range of ±1 m to ±4 m, depending on used services [10]. The accuracy of these services has a corresponding cost. Furthermore, the digital maps and the interfaces need to be improved to provide sufficient details for navigation by the blind and visually impaired people.

As pointed out in Sect. 3.3, advances targeting similar application areas for use by the general population, such as autonomous vehicles, can also be of interest to mobility-related AT. Tests of new software-based GPS have shown that they can improve the accuracy of locating a car to within 10 cm [11]. The software-based GPS with accuracy and low-latency can overcome the main three issues with using cameras and LIDAR in autonomous vehicles, namely inclement weather, blurred lane markings, and over the horizon blind spots. The proposed system requires data from ground base stations that need to be spaced within 20 km of each other. The processing and other hardware requirements make it impractical for use with the smartphones currently in the market. So, this accurate GPS may not be very suitable for autonomous vehicles. Its combination with other GPS-based ETA, however, may be of value for use in some AT for navigations. For example, real-time assistance prototype uses stereo cameras on a helmet, a laptop, and small stereo headphones to provide rich information for navigation [12]. HFVE is another complex system for creating and accessing audio-tactile images [13]. As with many other more technical approaches, for the reasons outlined, the adoption and acceptance of high-tech assistive products remain uncertain.

Besides GPS technology, there are other potential approaches for obtaining position data. The use of mobile phone signals, for instance, is one. Localization of a mobile phone may be done through its GPS or using trilateration of radio signals between the cell antenna towers and the phone. The trilateration approach can be based on the roaming signals emitted from the phone to contact the cell towers, without requiring an active or ongoing call. The legislation in the US and some other countries require that the location of a mobile phone that makes a call to specific emergency numbers be identifiable. The accuracy of positioning through network-based techniques, however, varies substantially from meters to tens of meters, even in areas with high base station density [14]. Either way, with GPS or use of radio signals, the information must be projected on digital maps to be related to the landmarks of interest.

As discussed in Sect. 3.3, position data can also be obtained using trilateration concepts with a radio signal from other communication technologies, including Wi-Fi, Bluetooth, UWB, and in some cases even NFC. Each is exhibiting their own advantages and shortcomings. In general, to achieve this, the signal beacons are generated within the environment and trilateration between them and a user device, can identify their location. The accuracy depends on the technology used and on the propagation effects of the signals in the environment.

Another approach that uses beacon signals to provide some ambient intelligence in a more straight-forward manner relies on the embedding of the navigational beacon generators in strategic locations of the environment. Depending on the individual's needs, they can be activated to provide either close-range indications for doorways, crossings, and bus stops, as well as longer-range information about directions to places of interest, such as buildings and public transport stations [10]. The handheld device needs to provide suitable interfaces with tactile and audio outputs. Talking signs, discussed in Sect. 4.4, are examples of such systems that provide directional voice messages transmitted to the device of a user by infrared light. Most of these developments are still works in progress, requiring more research and development efforts. A scheme that also relies on beacon sounds to guide the user along the required path is System for Wearable Audio Navigation [15]. It comprises of an audio-only output and tactile input. It is also capable of indicating the locations that may be of interest to the user in their proximity.

Ideas from self-driving cars can also provide radically different solutions for assistive navigation or recognition technologies, as well as for interacting with other machines and humans or interfacing. For example, Drive.ai is using AI and deep learning to develop systems that can be trained to interpret data from sensors and control a vehicle accounting for human behavior and interactions (https://www.technologyreview.com/s/602267/new-self-driving-car-tells-pedestrians-when-its-safe-to-cross-the-street/). For these systems, cameras are seen as available and cost effective sensors that may be replaced by solid-state LIDARs, depending on their availability and price. It is interesting to note that for autonomous cars, many experts believe that the perceptions of the environment and the surroundings cannot be based on traditional computer vision concepts. The major industry players, use some combination of LIDARs and other advanced sensors, HD maps, or deep learning for gaining awareness about the environment. Google, for example, has been pursuing deep learning for perception since 2012 (https://research.google.com/teams/brain/). The autonomous cars have many of the hallmarks of advanced IoT-based AT. They have some mechanical function, driving. Then, there are human-machine and machine-machine interactions. Humans can be the user of the car, drivers of other vehicles, or pedestrians. As such, based on very similar assertions to those made for AT developments in this work, some have argued that the self-driving car developments must be based on holistic approaches, rather than on modular tactics [16].

One aspect that most AT share with the autonomous vehicles relates to the human effects and the ways that cars, robots, or machines interact and exchange information with people. The interfaces and how the interactions are managed can make or break assistive products and their adoptions. Consider that using ambient intelligence, there is some information that may be specifically of interest to visually impaired or blind individuals. The information about their surroundings can be passed to them using smartphone apps. For example, BlindSquare can describe the location, announce points of interest and street intersections, as an individual makes a journey or moves around (http://blindsquare.com/). BlindSquare is an accessible app that uses GPS capabilities of iPhone or iPad to determine the individual's device location. It can then access other apps like Foursquare (https://foursquare.com/) and Open Street Map (http://www.openstreetmap.org/) for information about the surroundings of that location. FourSquare is a popular mainstream app that is based on using location intelligence and offers technology that is used in assembling many other location-aware and context-smart apps. BlindSquare can then determine the information that may be of interest to the user and speak it in a clear synthetic voice, available in many languages.

Face recognition, in a more general sense, can also be used in many care-related applications and AT. For instance, recognizing people that one can socialize with and respond to their presence, requires recognizing them. Face recognition technology is very mature and is already in use in various settings. Adopting the face recognition technology intended for other objectives, for AT development purposes, can provide many innovative solutions. For example, the good accuracy and speed of the face recognition technology FindFace that in less than 1 s can search through one billion photographs [17], or that of Google Image Search [18], or others can help with recognizing people known to an individual, using smartphones or any other Internet-connected devices. Google claims it can identify faces with a 99.6% accuracy with FaceNet algorithm, achieving 95.1% accuracy on YouTube [19]. Microsoft has patented a billboard that can identify a person as they walk by it [20]. Adaptation of these technologies to AT can be of great assistance, for instance to those living with mild dementia as it can remind them of familiar people around them. They can also be of value to visually impaired people who may be interested, for instance, if a familiar face is near them. Many smartphones can use these apps and achieve these recognitions.

The aged people have a strong sense of belonging to their place of living. This sense is related to a strong awareness of this space and what they can expect, which they have developed over many years. It is long established that such an awareness is part of their cognitive organizations [21]. Changes in residential places or care-givers, or introduction of AT for older adults can often cause conflicts with such a strong awareness and cognitive organization. The need to carefully assess the impli-cations of such conflicts, particularly for someone living with dementia, regarding their social, mental, and physical wellbeing and safety have been studied and well-reported [22]. An example that has been reported to reasonably adapt technology to the sensitivities of its users is the Just Checking system (http://www.justchecking.com.au/). This system is based on motion sensors that are installed in various areas of the user's place of residence or on the doors. A family member can set alerts or visit a protected website to see the activations of the sensors. Based on this informa-tion, they can get a quick and revealing representation of the overall activities of their elderly relative, and identify potential issues, such as the person not having gone to the kitchen for food and the like [23].

6.2 Interaction Requirements

Many IoT-based products and smart devices are finding their ways into the lives of people. It appears that the devices like smartphones that are easy to interact with, penetrate our lives more extensively. Assistive technologies can be part of this same pattern of adoption and employment. Moreover, the acceptability and usability of AT will be dependent on the practicality of the technology-human interfaces for employment by the elderly or individuals living with disability or dementia. The advanced IoT-based AT and assistive environments have capabilities of deploying interfaces that facilitate more natural human-like interactions. They may include the use of natural language and spoken words, touch screens, gestures and emotional intelligence, and haptic feedback.

Human interactions rely mostly on vision and sound, and to a lesser extent on body language and gestures, and then on touch and other mechanisms. To provide natural experiences and extend on them, smart assistive devices and environments need to employ similar interaction mechanisms. For instance, compared to key-boards, voice controlled devices can be seen as providing more user-friendly inter-faces that are easier to operate for a visually impaired individual or an older adult. On the other hand, for a person with hearing impairment, interactions through a touch screen may be more convenient than using voice commands.

As discussed in Sect. 3.2, voice recognition and consequently voice to text con-versions are very mature fields. Many systems that are available for smartphones, tablets, and stronger computing platforms provide rather accurate and easy to use voice recognition capabilities. Deep learning approaches have significantly contributed to natural language processing in effective manners. Most newer Android and iOS based smartphones incorporate voice recognition into their native apps and facilitate their use by app developers. These features meeting the universal design concepts, usable by all people operating these phones, can also provide assistance to individuals that due to some impairment may require them. The large market of the general population can also foster more developments that can be of value for implementations as interfaces in AT.

Vision interactions as a way of controlling a device or some app is not as effective as voice commands. For most humans, vision is a natural method of gaining information, rather than controlling a device. As such, equipping smart devices and environment with cameras, LIDARs, or other sensors that acquire the information that people usually gain through sight, can provide interesting solutions for various situations. For instance, the interfaces can be used to facilitate gesture control [24]. They can also be used for tracking and localization of humans or other objects [25].

The ease or difficulty of using interfaces has a direct relation to the cognitive load and effectiveness of the device or service that incorporates them [26]. For any AT, the cognitive load is an overhead that must be reduced as much as possible. Reducing this load is of particular importance for navigation systems assisting the blind or visually impaired people or the elderly. Several studies have been carried out that indicate the relationship of the cognitive load with the effectiveness of using auditory, haptic, vibration, and kinesthetic cues [27]. While these studies cannot quantify the load, they have shown that some of the advanced technologies, such as digital maps that are used on a touch screen device, can be cognitively demanding tasks. This inference is in line with the discussions in Sect. 5.2, that outlines several other reasons leading to low adoption and acceptability of the more advanced AT.

With the pervasiveness and usefulness of smartphones, their interfaces deserve particular attention. Many commercially available mobile devices are equipped with uniform touch-based input and auditory-tactile output. As such, some of the big players of the mobile platforms are, perhaps indirectly, setting the scene for the standards for the interfaces of the AT [12]. Voice commands and use of natural language are quickly finding their ways into the human-smartphone interactions. However, the human interactions with smartphones are still predominantly based on vision and use of the touchscreen to tap an icon or to type on an onscreen keyboard. These types of interactions can put some people, particularly the blind or visually impaired at a disadvantage, making smartphone inaccessible to them. Braille has been successfully used by visually impaired people for nearly two centuries as a system of reading and writing. There are many applications for touchscreen devices that offer Braille, using virtual keys. Considering the normal sizes of screens on smartphones, the required six-finger Braille typing is not deemed to provide a good solution, particularly, considering that the user needs to put their phone on a surface while typing with both hands. One-handed Brailling has been proposed as a way of overcoming such issues and allowing a visually impaired person to comfortably type with only one hand on a smartphone or a smartwatch that supports multi-touch screen [28]. The proposed system employs an adaptive calibration scheme that allows the user move their hand while typing.

There have also been solutions that rely on native or developed apps for the smartphones. Apple, one of the main market players, has provided several popular solutions that facilitate the use of Braille on smartphones (http://www.apple.com/accessibility/). Among them, a native accessibility feature that works with VoiceOver is the standard QWERTY keyboard. To use the keys, the user moves their finger over the virtual keyboard, receiving audio feedback for the touched character that can be selected through other gestures. There are several other two-handed Brailling schemes, including BrailleTap, BrailleType, and BrailleTouch, as well as one-handed systems like TypeInBraille and Perkinput, each with their own advantages and shortcomings [28].

The iOS also provides some applications for navigation and transport. Apple supports proper interfaces that make these apps capable of providing assistance to the visually impaired people. The Maps is one such app that can be used with most iOS platforms and is accessible using VoiceOver (http://www.apple.com/ios/maps/). It can produce turn by turn spoken instructions for getting to a set destination. A user can also use third-party apps inside the Maps. For example, Ariadne GPS works with VoiceOver seamlessly (http://www.ariadnegps.eu/). With this app, a visually impaired person can identify their position and track it as they walk. They can choose to be informed of street numbers and names. It also has the options for exploring nearby settings. Working with the VoiceOver, it can alert the user of approaching the points of interest using sound, vibration, and voice. The current version of this GPS can work anywhere that Google Maps are available. Several other apps work with iOS or Android and provide similar functionalities [12].

The interfaces and interaction mechanisms can also be of vital importance for people with varying abilities and the elderly. In fact, several studies have shown that usefulness and ease of use are the two most important factors foretelling the technology acceptance by the older persons [29–31]. Taking that into account, some systems have been developed in a more customized fashion to make human-machine interactions easier. For instance, a custom suite of AT applications for Android, named Universal Gateway for Android (UGA), provides several interfaces for computer and smartphone users [32]. This Linux-based embedded system is capable of speech recognition, intended for people who use voice as their primary means of interaction for access to computers. This functionality can be powered by Google speech recognition services, Cloud Speech API (https://cloud.google.com/speech/). UGA can utilize motion sensors of a smartphone and interpret the detected movements as dislocations intended for the mouse pointer. The system also provides individualized virtual keyboards, an app that enables the use of the smartphone touchscreen as a touchpad for computer devices, and creation of a virtual push button on the screen of a phone.

For many wearables and IoT objects, the interfaces can be the cause of serious concern. Smartwatches, for instance, have very small touchscreens, making it necessary to have interaction mechanisms that rely on voice, haptic, gestural interfaces, or communications through another device like a smartphone. Voice commands seem to be reliably used in some brands, such as Android Wear, while others are pursuing interaction through the accelerometer or providing haptic feedback [33].

Many IoT-based devices, like some smart appliances or medical equipment, lack a screen or some other convenient interface to facilitate interaction with their users [34]. They may have no clear way to notify the user when a problem arises.

Evaluation of how well an AT and its interfaces work can pose serious difficulties. It can be especially difficult, when growing areas, such as the IoT, also need to be incorporated. Considering the definition of AT, mentioned in Sect. 3.1, it covers a diverse range of devices, applications, and services. The expected functional outcomes also cover an extensive array of assistive capabilities. The usability and acceptability by an individual, which can be one of the decisive factors in employing a device, can relate to the need for customized applications, pointing to the further potential diversity of the AT. For some situations, a user may prefer to have one device that can assist them in several activities. While for another case, a person may find such a device too complicated, too expensive, or unnecessary and therefore, prefer having more than one device.

Some categorization that may help with having guidelines to evaluate how a particular approach or technology is working for the elderly or people with disability can be quite helpful. Some of the points that can be considered in that context include the purpose and functionality, technology type, interactivity and interaction mechanisms, and portability [10]. Other criteria that can be included are availability, affordability and cost, aesthetics and looks, social and cultural acceptability, ease of use, and universal design conformance.

Organizations working for the elderly or people with disability, the affected individuals, and their caregivers, as well as the AT designers and manufacturers and those who coordinate services or provide therapies, can all contribute to the aged care and disability experiences. As with many other cases, user involvement is gaining importance in evaluating the quality of the AT. Such participation, can empower the intended users and put the spotlight on their needs based on their experiences [35]. Active involvement of users can also provide the means for practical, real-world evaluations that are not centered around extrapolation of some lab-based experiment. The assessment of forthcoming AT, particularly those relying on advanced technologies like IoT, must go beyond whether it delivers on some initial functionality or not. The challenge for the evaluations is that they must be based on AT interactions with all levels of the user experience [36].

In many areas, high-tech assistive devices and services are available or proposed. However, as the discussions in Sect. 5.2 indicated, their adoption and employment do not come even close to the use of many simple traditional AT. To improve on what widely used traditional AT can do, perhaps with more advanced technology, several points need to be considered. In some cases, developments that can assist the elderly or people with disability can be of value to the population as a whole. In other cases, the expenses or the assistive components, like the audible alarms, can be of concern or annoyance to some people. These cases may simply correspond to poor design or improper implementation of the assistive components. For both cases, though, implementations of "design for all" concepts can be of value in reaching the right solutions. The use of technology must be justified, based on the assistance that it provides, rather than on being high tech or new. Expenses, ease of use, availability, languages used are among the factors that play significant roles in evaluating an AT solution.

Besides considering all of these aspects that mandate taking a holistic view of AT developments, it is also imperative to base that view on the inclusion of all stakeholders. It is important to note that the evaluation of a device or service may produce dissimilar outcomes for different stakeholders. For example, policymakers may find cost-savings associated with some technology a positive outcome, while the users or the caregivers may not be focused on that aspect at all. Another point that makes taking a holistic view important, relates to relevancy and rigor of studies carried out by different stakeholders. Some rigorous lab-based studies are dismissed as irrelevant to the real world considerations, while those that have attempted to use practical experiments are usually very limited and can hardly be generalized [36]. A comprehensive literature review has also identified that the use of different terminology has restricted collaborations among the professionals and researchers [37]. This review has also indicated that concentration of researchers and developers on their own areas of expertise, without sharing or consulting with professionals in other areas working on the same types of problems, has prevented providing the users with a consistent set of assistive solutions and technology.

As with some other aspect of the developments of IoT-based AT, the advancement of their interfaces may benefit from progress made in other fields. For example, the interactions of self-driving cars with their human controller, other vehicles, and people can be relevant. Some projects are aiming to replicate human emotional intelligence and benefit from its management of social behavior and non-verbal communications to improve the interactions of an autonomous vehicle with other cars and pedestrians (http://www.reuters.com/article/drive-ai-autonomous-idUSL8N1B040J). This project aims to enable the car to show intent and make complex interactions possible, making it clear if the car does see a pedestrian or a motorcycle, making it safe for them to pass it, for instance. Some aspects of these interactions relate to social impacts and acceptability of the smart autonomous vehicles. Such aspects can also significantly influence the adoption and employment of assistive IoT-based devices, technology, and environments. The influences of social interactions on AT usability are considered in the next section.

6.3   Societal Impacts and Social Usability

Various social and personal issues affect the experiences of the elderly and people with disability or dementia. These experiences, which are also influenced by legislations, policies, design and aesthetics of the products, and awareness regarding the available devices and services make people accept or abandon a given AT. Many studies have already established that most affected individuals are aware that AT cannot fully replace sight, hearing, cognitive capacities, or any functioning [5]. They have also indicated that what is valued is access to information and services like everyone else in a fashion that employing assistive or mainstream devices lets them be seen as capable as anyone else. However, many individuals and their families may not be well-informed of the available services, or feel disempowered, or appear unable or unwilling to express their needs [35]. In most countries with advanced services, strong organizations of persons with disability or elderly have driven the changes through lobbying governments for reforms.

Equal rights and access for people with disability or senior citizens have had similar footings as other Civil Rights Movements of the 1960s in the United States [5]. People with various abilities rallied to reject the notion of disability being essentially a medical condition and instead asked for legislation and policies to be based on the more positive and empowering views of social models that consider access or disability the result of situational and societal constructs. While the ensuing reforms and the resources that go with them have been positive, stigma and negative attitudes of others, can also greatly influence the feeling of empowerment and subsequently the employment of the AT.

A survey of AT literature has identified three major reasons for abandoning of assistive devices [38]. The reasons include, the AT failing to accept an individual's disability, the device made the user stand out from others resulting in social exclusion, or the AT caused the exclusion of the person due to clashes with socially accepted cultural values. As such, it is not merely useful functionality that results in the widespread adoption of the AT. The functionalities and other complex factors make a device or service accepted, as fit for purpose or not, by their intended users.

An obvious first step for any technology development, including for the AT, is to clarify its purpose. The medical and the social models of disability, explored in Sect. 2.3, provide some answers to this probe. The focus of the medical view is on improving the functional capabilities of the person, while the social model aims for the removal of environmental and societal barriers. Obviously, one can also argue that a more combinatory view needs to guide the AT developments [36]. The lack of holistic, integrative, and interactional views may result in ignoring the complexities of the experiences of the affected people, making the AT unacceptable to their intended users. The main purpose of AT is perhaps to enable their users to have equal access to information and services and participate in the opportunities and activities that the society offers, just like everyone else [5]. As such, even when the AT provide the right functionalities, if they do not provide equal access or worse, result in stigma and exclusions, it has failed to fulfill its purpose.

While many AT users may feel empowered when using their devices, they may not accept a device if it has a possibility of misrepresenting them or characterize who they are in the wrong or unfavorable ways. For instance, a young person with hearing loss may feel disinclined to wear hearing aids, as the impairment is often associated with the elderly. A study of the AT for dementia found that none of its participants considered the Disabled Living Foundation as a source of information, perhaps due to an objection to the word "disabled," or that the carers did not consider dementia as a disability [39]. All people associate some characteristic to the objects they or others often use. On occasions, these make up our first impressions of others and can form our unconscious judgments and expectations of their abilities and behavior [40]. For a person using an AT, such first impressions may result in feeling self-conscious about standing out from others and their abilities being misjudged. The ensuing biases may be related to the misconception that an assistive device implies that its user needs continual assistance.

For some individuals in some specific situations, the feeling of being misjudged about their abilities may be more influential than the functions provided by a device, service, or environment. For some others, it may be the other way around. Some assistive devices, such as a white cane, have strong social meanings associated with them. The difficulty arises from misconceptions that may also be associated with them. Wrong assumptions, social norms, and occasionally ignorance result in misconceptions permeating the use of AT. Two misperceptions that have been cited in the literature are that AT can functionally eliminate a disability and that people with disability are helpless without their devices [5]. Misconceptions, particularly of the second kind, can be particularly important for an individual in employing an AT or abandoning it. As the individual may find it important for their confidence or for other reasons that other people do not perceive them as incapable of doing various tasks.

The perception of others about the capabilities of an individual have significant ramifications for all of us, particularly in situations like seeking employment or changing jobs. The misperceptions associated with some AT employment, can sway some of these users into abandoning their assistive device if possible at all. Such an abandonment can be particularly the case, when an impairment may not be obviously evident by itself. In situations like this, a person may choose not to disclose their impairment. Consequently, they will not use technologies and devices that can make them stand out or give away their impairment. Alternatively, if the assistive devices looked comparable to what is used by many people, for example like a smartphone, they will be more likely to be taken up. There can, of course, be AT users who may choose to draw attention to their disability, for instance, to access a service, and they may do so through their devices. Some may prefer to assure others of their capabilities and independence facilitated by their AT. Either way, AT acceptability, has many complex factors, including functionality and usability, perception and misperception, stigma, and aesthetics [4].

The design and aesthetics of the assistive devices can tremendously affect their acceptability. The devices that have successfully addressed this issue, for example, the eyeglasses and sunglasses, have found their way into being used by people with all types of abilities. This issue seems to have been overlooked by many developers and manufacturers. Obviously, the design for all ideas, explored in Sect. 2.3, can also provide efficient and attractive solutions. Such devices are intended to be used by different people, irrespective of their age, gender, health status, and other factors.

The Society also affects the age and disability experiences through other means. These include the regulatory frameworks that are discussed in the next section. The perception of the society as a whole and the way community looks at the elderly or people with disability manifest themselves in the forms of policies. Until the 1960s, people with cognitive and many sensory impairments were segregated in residential and educational institutions [35]. The separation was a widely accepted practice by most communities in the developed countries, which did not facilitate the participation of these individuals in the society and did not let them have much control over their lives. With changes in attitudes, many studies have shown improved quality of life for people in community care [41, 42]. It is well-established that community-based care provides much better services and better quality of life, in much more cost-effective manners compared to institutional care [35]. Smart environments can be the facilitators of such setting, providing extra benefits that have been outlined in this work.

All of these considerations point to the serious impacts of the society on the age and disability experiences, as well as on how they are approached, and on the use and acceptability of AT. Related policies and the developments of AT must aim to lessen the social misperceptions about age and disability and alleviate the effects of misconceptions on individuals. Development of devices and services that work with mainstream technologies or make them more accessible can be of value for this aim. Such developments can be part of a holistic approach to assistive IoT and smart environments that incorporate notions of social acceptance, among many others that have been considered in this work so far. These will be further discussed in the next chapter. Another important aspect, regulatory frameworks and their impacts on the AT and aged care, will be the subject of the next section.

6.4   Regulatory Frameworks and Considerations

The positive and empowering views of the social disability model, discussed in Sect. 2.2, and the demands of the aging populations, have led to the development of civil rights laws and regulatory reform for the elderly and individuals with disability. While these have led to solid progress in some countries and for some situations, people with disability and the seniors remain vulnerable, as they may depend on a large number of caregivers or have communication difficulties [35]. The vulnerabilities in an institutional and to a lesser extent in a community-based care can put a person at the risk of isolation, lacking stimulation, or even physical and sexual abuse. There are also the accessibility and inclusion issues that need to be addressed. To counteract these risks, voluntary endeavors have had some results, which were neither widespread nor sufficient. The need for safeguards and provisions of equal rights and access have led to many significant national and international laws and regulatory frameworks that are briefly considered in this section. After some general discussions relevant to these frameworks, in line with the theme of this work, the focus of this segment will turn to legislation that relates to ICT and IoT-based AT.

Perhaps, one of the earliest rules that deal with the rights of people with an impairment is the one in the Bible (Lev. ch 19, ver 14) forbidding that stumbling blocks be left around, as they could cause blind people to fall [10]. Since then, there have been many legislations making it illegal to discriminate based on age or disability. Some have also set the scene for accessibility to buildings, facilities, and services which are the cornerstones for inclusion. For instance, in the US, mandatory minimum standards for accessibility were passed in 1968, after it became apparent that the voluntary standards set in 1961 did not achieve their aims [43]. Many other countries now have similar legislation. They mostly combine antidiscrimination and the accessibility standards to infrastructure and facilities in their regulatory frameworks. However, according to the UN, in 2006 only 45 countries had antidiscrimination and other disability-specific laws [44]. The implementation levels and efficacy of those legislations have also been criticized by many disability support organizations [10]. One of the serious international attempts to change these situations, specifically for people living with disability, whether young or old, resulted in the adoption of CRPD in December of 2006 [45].

CRPD has so far been ratified by 173 countries, carrying the force of their national legislations [45]. It encourages the full integration of persons with disability in societies. Many other international documents have stressed the need for managing disability issues as a human rights matter. These include the World Programme of Action Concerning Disabled People (1982), the Convention on the Rights of the Child (1989), and the Standard Rules on the Equalization of Opportunities for People with Disabilities (1993). However, their adoptions by countries around the globe are nowhere near that of the CRPD, which outlines the human rights as well as the civil, cultural, political, social, and economic rights of persons with disability [35].

The Convention signatories are required to identify and eliminate all physical obstacles and barriers and ensure that persons with disability can access various facilities and services, including ICT. It also fosters the independent living of individuals and mandates the provision of services to facilitate their mobility through affordable access to the required AT. It includes provisions for equal access to education and other important aspects of life. Its Article 21 is particularly relevant to this work. It asks the countries to provide information in accessible formats and technologies. Article 30 aims to ensure that television broadcasts, films, and similar material are in accessible formats to foster inclusion and participation (http://www.un.org/disabilities/convention/convention.shtml).

In the US, the enactments of Section 504 of the Rehabilitation Act of 1973 and the Americans with Disabilities Act (ADA) in 1990 provided many practical protections for individuals with disability. These protections included mandating that education institutions remove barriers, eliminate discriminations, and facilitate inclusion of students with disability [46]. They also make provision of accommodations in public spaces obligatory, but without requiring retrofitting [47]. As such, the spaces that were built before the enactment of the legislation may remain inaccessible.

Similar antidiscrimination laws have also been passed in many other developed countries. These laws include the UK Disability Discrimination Act (1995), Australia's Disability Discrimination Act (1992), Canada (1986, 1995) and the 2005 Canadian act of Accessibility for Ontarians with Disabilities, and New Zealand (1993) and the Disability Strategy of 2001. Some other countries, like Germany and South Africa, have included related clauses into their more general legislation [35]. In general, the purpose of the legislations is not to ensure the success of a person with a disability. They rather aim to provide equal opportunities for people with any abilities. These legislations, however, have various characteristics. Some of them, like those in Germany, introduce affirmative actions, while others, as in Australia and Sweden, include formal services that are part of a national disability policy, aiming to improve community participation of the affected individuals.

The challenge for these laws and regulations is that the economic and financial motivations can result in overlooking the needs of the diverse population. As such, their effectiveness and benefits in providing access and safeguarding against discrimination as a result of different ability, age, or other factors of a similar nature can only be realized if they are ingrained into applicable policies and practices [48]. If the enforcement processes involve taking legal actions by people with disability or similar mechanisms, the laws need to be redesigned, as lawsuits can be expensive and exhausting, requiring resources that are not available in many cases [35].

The other topic relevant to the effectiveness of the regulatory framework relates to the definitions and conditions that enable an individual to benefit from these laws and policies directly. For instance, the US federal ADA definitions consider a disability as a "substantial limitation" of a major life activity [46]. Some states, like California, drop the word "substantial" making it easier for more people to access related accommodations and services. The federal legislation considers major life activities to include breathing, walking, talking, hearing, seeing, eating, learning, reading, concentrating, and thinking. While these are broad, impairments of major bodily functions and systems are expressly included in the 2008 ADA Amendments Act. As such, disability protections of the ADA clearly cover individuals with chronic health conditions or diseases, such as diabetes or cancer. Such protections can relate to the provision of services for many older adults. While traditionally the disability assessments were mostly based on the medical model and criteria that related to it, the focus of the recent past has been on meeting the requirements for efficient functioning and being active. The new view, stemming from the social disability model, is a widely accepted one that is based on the ICF [49].

The third issue relevant to the regulatory frameworks relates to the inclusion of the elderly and people with disability, as well as their families and their informal carers in making decisions about the services they use. Making providers accountable to consumers has been shown to improve the service outcomes [50]. This inclusion can also empower the individuals and result in more personalized services. Many developed countries, including Australia, Canada, several European states, and the US have been replacing the generic services with the more flexible systems that may incorporate individualized funding models [35]. In these models, based on need assessments, public funds are allocated to individuals and are controlled by them, with constraints on paying for AT, therapies, or other approved expenses. Professional are also available for assisting the consumers when needed.

With regards to the role of AT and its usage in improving the quality of life and inclusion of individuals, the Technology-Related Assistance for Individuals with Disabilities Act of 1988 in the US has been a turning point. It drew attention on how AT can positively affect the educational outcomes, work performances, and social life of its users. The Disabled Student Allowance of 1993 in the UK and the US

Individuals with Disabilities Education Act of 1997 have also provided many opportunities for education-related AT research and development [51].

ICT accessibility and provisions of universal services have been mandated in regulatory frameworks in many developed countries. Australia, Canada, and the US have pioneered many ICT accessibility policy changes that have been used to set the standards in several other countries [35]. For example, in many countries, laws and regulations mandating the provisions of accessibility to films and videos, have been passed. However, in most cases, these provisions may not cover media on computers, smartphones, or other portable devices. The US Television Circuitry Decoder Act makes it compulsory for the TV set manufacturers to include technology for supporting closed captioning [52]. In Australia, captions for all television programs broadcast between 6 PM and midnight, as well as for all news and current affairs programs on the primary channels of all free-to-air networks must be available (http://www.mediaaccess.org.au/tv-video/captions-on-tv/captions-on-digital-tv). In Japan, the target is that by 2017, all live and prerecorded programs must have captioning available if technically possible [35]. In Denmark, the Act on Radio and Television Broadcasting of 2000 makes subtitling of public service television channels mandatory. Ireland, Italy, Finland, and Portugal news programs are broadcast with sign language interpretations [53].

The regulatory framework, particularly the parts related to technology and AT, can be out of pace with the trends of the advances and service developments. In the US, while the Telecommunications Act of 1996 regulated many basic services, including the telephony and television, it left enhanced services, such as the Internet, unregulated [54]. Many Internet-based multimedia services are catching up with the traditional media, like television or DVD. However, without strong regulations, the rapid pace of new network-based services can on occasions ignore the accessibility requirements.

The US has addressed some of the issues relevant to enabling people with disability to access broadband, digital, and mobile innovations through the Twenty-First Century Communications and Video Accessibility Act of 2010 (https://www.fcc.gov/general/twenty-first-century-communications-and-video-accessibility-act-0). Some of its provisions require that television equipment, like the set-top boxes, be made accessible and usable by people with vision loss or other impairments. These provisions, for instance, make it easier to select and use Secondary Audio Program (SAP) channels that explain a program in different languages (http://www.afb.org/info/living-with-vision-loss/using-technology/video-description/accessing-video-description/1235). In some countries, legal challenges have led to improvements in provisions of accessible telecommunications services. In Australia, for instance, the rulings in the case of Scott and DPI v. Telstra resulted in telecommunications access being defined as a human right [55].

Regulatory frameworks covering ICT issues are in force in many countries. However, their coverage relevant to accessible ICT still leaves much room for improvements. The CRPD, which mandates the development of national ICT accessibility policies, by more than 173 countries that have ratified it, has helped. By 2014, over 50% of these countries had expanded the definition of accessibility in their regulatory frameworks to include computers, video programming, cellular systems, the Internet, and other ICT-based AT (http://www.g3ict.org). In many developed countries, the laws and regulations that are in force can easily fall behind the rapid technological advancements. In some cases, to allow for those advancements to take place, the regulatory bodies may simply overlook the accessibility needs. Gaps in many devices and services that are part of everyday life for many people make them unusable for the elderly or people with disability. These include accessibility issues with some websites, smartphones and their related apps, some television equipment and programs, and many IoT-based devices. The diversity and prevalence of ICT and the services that are provided through them may actually make it impossible or impractical to legislate for all potential cases. For example, access to telephony is well regulated for landlines and in some case even for mobile and cellular systems. However, provisions of access to Voice over IP, or telephony over the Internet, are not usually regulated [35]. To address ICT accessibility, smarter regulatory frameworks are needed.

Some rather extensive studies in Europe have indicated that countries with robust legislation and regulatory frameworks accomplish higher levels of ICT access (https://ec.europa.eu/digital-single-market/en/news/assessment-status-eaccessibility-europe). The governments that have passed legislations relevant to accessible ICT have used different approaches with varying degrees of comprehensiveness and details. In the US, Section 508 of the United States Rehabilitation Act includes procurement policies that require accessible equipment. Such procurement policies can also provide incentives for the industry to manufacture equipment that adopts standards for universally designed technology [35]. The UK and the EU have also passed many laws and resolutions on ICT public procurement policies, as well as on other related issues, such as web accessibility for the elderly and people with disability [56, 57]. Strengthening ICT accessibility is among the major aspects of the European Action Plan, passed as part of the EU initiatives to meet CRPD requirements [58].

For the IoT-related areas, the majority of the regulatory developments are towards the scope of applications or managing the privacy and security issues. Both of these topics are of vital importance for IoT-based AT developments and adoptions. Many new applications are coming up, as well as many discussions on how the IoT regulations should go. However, the need for the IoT-specific legislation is still a highly debated topic.

One of the most comprehensive and well-cited surveys on this subject, is the one conducted by the European Commission in 2012, the results of which were published as the "Report on the Public Consultation on IoT Governance" in 2013 (https://ec.europa.eu/digital-single-market/en/news/conclusions-internet-things-public-consultation). With the diversity of expressed opinions, the European Commission at the time, essentially abandoned its active efforts relevant to the regulatory framework of the IoT, and instead, focused on privacy, security, and trust issues [59]. Even with that specific focus on security and personal safety in the IoT, many divergent views were expressed. Major industry players, while acknowledging the need for public guidelines, were against any additional regulations [60].

In September 2014, the Article 29 Data Protection Working Party of the European Commission issued its opinion on Recent Developments on the Internet of Things (http://ec.europa.eu/justice/data-protection/article-29/documentation/opinion-recommendation/files/2014/wp223_en.pdf). The document does not reach any solid conclusions on the need for IoT regulatory issues. However, it acknowledges that some developments of the IoT need to be controlled. Otherwise, they may "develop a form of surveillance of individuals that might be considered as unlawful under EU laws." By noting that the IoT devices "collect and further process the individual's data," relevant existing EU Laws can be invoked, where most of the security and privacy burden is put on the shoulders of various stakeholders.

The European Commission established Alliance for Internet of Things Innovation (AIOTI) in March 2015 with the aim of creating 2020 European IoT roadmap (https://ec.europa.eu/digital-single-market/alliance-internet-things-innovation-aioti). The relevant Working Group of the AIOTI, has also concluded that there is no merit in enacting IoT-specific regulations, at this stage. The Group has given two main reasons for reaching this conclusion. They note the high risk of inefficient and error-prone regulatory frameworks in the complex and fast-progressing environments, such as that of the IoT. The second reason is based on the observation and belief that the existing laws and self-regulations can efficiently address breaches and failures, while any new legislation should only focus on "well-defined market failures" [60].

In November 2013, the US Federal Trade Commission (FTC) held a workshop to explore privacy and security issues relevant to the IoT. A Staff Report, summarizing the main outcomes and views of participants and Commission staff, was subsequently released in January 2015 [34]. The Report acknowledges the data privacy and security threats that can only intensify by the emerging IoT technologies. It also identifies various privacy risks that can be associated with the IoT. The risks include a direct collection of sensitive personal information, which are similar to those present in other Internet and mobile commerce services. Through a collection of data on personal habits, health conditions, and locations, especially over time, the IoT-based systems pose the risk of enabling the inference of sensitive information by unauthorized parties. It notes that current legal protections, such as those provided by the Fair Credit Reporting Act, may not be enough even if some companies use this information to make credit, insurance, employment, or similar decisions. The Commission staff, considering the "great potential for innovation" in the IoT technology and its related areas, reiterated its previous recommendation for the "US Congress to enact strong, flexible, and technology-neutral federal legislation to strengthen its existing data security enforcement tools and to provide notification to consumers when there is a security breach." The Commission staff saw the solution lies in broad-based, rather than IoT-specific, privacy and data security legislation and self-regulation by the industry.

It can be noted that even in countries that laws and regulations dealing with assistive ICT and high-tech devices and services have been passed, these regulations have occurred with quite substantial lags to the widespread use of the underlying technology or service by the public. Assistive IoT and smart environments do not appear to be any exceptions to this rule. As the need for the IoT-specific legislation is not yet established, regulatory frameworks relevant to assistive IoT and smart environments should be based on the broader existing laws and policies, at least for now and for the near future.

References

1. Schalock, R. L., Brown, I., Brown, R., Cummins, R. A., Felce, D., Matikka, L., & Parmenter, T. (2002). Conceptualization, measurement, and application of quality of life for persons with intellectual disabilities: Report of an international panel of experts. *Mental retardation*, *40*(6), 457–470.
2. Wang, M., Schalock, R. L., Verdugo, M. A., & Jenaro, C. (2010). Examining the factor structure and hierarchical nature of the quality of life construct. *American Journal on Intellectual and Developmental Disabilities*, *115*(3), 218–233.
3. Lancioni, G. E., & Singh, N. N. (2014). Assistive technologies for improving quality of life. In *Assistive Technologies for People with Diverse Abilities* (pp. 1–20). Springer New York.
4. Pullin, G. (2009). *Design meets disability*. MIT press.
5. Shinohara, K., & Wobbrock, J. O. (2011). In the shadow of misperception: assistive technology use and social interactions. In *Proceedings of the SIGCHI Conference on Human Factors in Computing Systems* (pp. 705–714).
6. Moritz, E. F., Biel, S., Burkhard, M., Erdt, S., Payá, J. G., Ganzarain, J., & Cabello, U. V. (2014). Functions: How we understood and realized functions of real importance to users. In *Assistive Technologies for the Interaction of the Elderly* (pp. 49–68). Springer International Publishing.
7. Xiao, J., Joseph, S. L., Zhang, X., Li, B., Li, X., & Zhang, J. (2015). An assistive navigation framework for the visually impaired. *IEEE Transactions on Human-Machine Systems*, *45*(5), 635–640.
8. Apostolopoulos, I., Fallah, N., Folmer, E., & Bekris, K. E. (2014). Integrated online localization and navigation for people with visual impairments using smart phones. *ACM Transactions on Interactive Intelligent Systems (TiiS)*, *3*(4), 21.
9. LeBrun, J., & Chuah, C. N. (2006). Bluetooth content distribution stations on public transit. In *Proceedings of the 1st international workshop on Decentralized resource sharing in mobile computing and networking* (pp. 63–65).
10. Hersh, M. A., & Johnson, M. A. (2008). Mobility: an overview. *Assistive technology for visually impaired and blind people*, 167–208. Springer London.
11. Ross, P. (2016). Centimeter-Level GPS Positioning for Cars. IEEE Spectrum. Available online: http://spectrum.ieee.org/cars-that-think/transportation/sensors/centimeterlevel-gps-positioning-for-cars
12. Csapó, Á., Wersényi, G., Nagy, H., & Stockman, T. (2015). A survey of assistive technologies and applications for blind users on mobile platforms: a review and foundation for research. *Journal on Multimodal User Interfaces*, *9*(4), 275–286.
13. Dewhurst, D. (2009). Accessing audiotactile images with HFVE silooet. In *International Conference on Haptic and Audio Interaction Design* (pp. 61–70). Springer Berlin Heidelberg.
14. Clarke, R. (2001). Person location and person tracking-Technologies, risks and policy implications. *Information Technology & People*, *14*(2), 206–231.
15. Walker, B., & Lindsay, J. (2006). Navigation Performance with a Virtual Auditory Display: Effects of Beacon Sound, Capture Radius, and Practice. *Human Factors, 48* (2), 265–78.
16. Ackerman, E. (2016). Drive.ai Solves Autonomous Cars' Communication Problem. IEEE Spectrum. Available online, http://spectrum.ieee.org/cars-that-think/transportation/self-driving/driveai-solves-autonomous-cars-communication-problem [Accessed: 10-Oct-2016].

17. Walker, S. (2016). Face recognition app taking Russia by storm may bring end to public anonymity. The Guardian. Available online, https://www.theguardian.com/technology/2016/may/17/findface-face-recognition-app-end-public-anonymity-vkontakte [Accessed: 10-Apr-2017].

18. Google, Google Images. (2016). Available online, https://www.google.ca/imghp?gws_rd=ssl [Accessed: 10-Oct-2016].

19. Schroff, F., Kalenichenko, D., & Philbin, J. (2015). Facenet: A unified embedding for face recognition and clustering. In *Proceedings of the IEEE Conference on Computer Vision and Pattern Recognition* (pp. 815–823).

20. Sobel, B. (2015). Facial recognition technology is everywhere. It may not be legal. The Washington Post. Availble online, https://www.washingtonpost.com/news/the-switch/wp/2015/06/11/facial-recognition-technology-is-everywhere-it-may-not-be-legal/?   utm_term=.f28bc89f7715 [Accessed: 10-Apr-2017].

21. Rowles, G. D. (1991). Beyond performance: Being in place as a component of occupational therapy. *American Journal of Occupational Therapy, 45*(3), 265–271.

22. Riikonen, M., Paavilainen, E., & Salo, H. (2013). Factors supporting the use of technology in daily life of home-living people with dementia. *Technology and Disability, 25*(4), 233–243.

23. ALM, N., & ARNOTT, J. (2015). Smart Houses and Uncomfortable Homes. *Studies in health technology and informatics, 217*, 146.

24. Garber, L. (2013). Gestural technology: Moving interfaces in a new direction [technology news]. *Computer, 46*(10), 22–25.

25. Li, R., Lu, B., & McDonald-Maier, K. D. (2015). Cognitive assisted living ambient system: a survey. *Digital Communications and Networks, 1*(4), 229–252.

26. Allen, G. L. (Ed.). (2007). *Applied spatial cognition: from research to cognitive technology.* Mahwah, NJ: Lawrence Erlbaum Associates.

27. Martinez-Sala, A. S., Losilla, F., Sánchez-Aarnoutse, J. C., & García-Haro, J. (2015). Design, implementation and evaluation of an indoor navigation system for visually impaired people. *Sensors, 15*(12), 32168–32187.

28. Šepić, B., Ghanem, A., & Vogel, S. (2014). BrailleEasy: One-handed Braille Keyboard for Smartphones. *Studies in health technology and informatics, 217*, 1030–1035.

29. Caprani, N., O'Connor, N. E., & Gurrin, C. (2012). Touch screens for the older user. In Assistive technologies. InTech.

30. Mitzner, T. L., Boron, J. B., Fausset, C. B., Adams, A. E., Charness, N., Czaja, S. J., & Sharit, J. (2010). Older adults talk technology: Technology usage and attitudes. *Computers in human behavior, 26*(6), 1710–1721.

31. Selwyn, N. (2004). The information aged: A qualitative study of older adults' use of information and communications technology. *Journal of Aging studies, 18*(4), 369–384.

32. Mulfari, D., Celesti, A., Fazio, M., Villari, M., & Puliafito, A. (2015). Embedded systems for supporting computer accessibility. *Studies in health technology and informatics, 217*, 378.

33. Rawassizadeh, R., Price, B. A., & Petre, M. (2015). Wearables: Has the age of smartwatches finally arrived? *Communications of the ACM, 58*(1), 45–47.

34. Federal Trade Commission. (2015). Internet of Things: Privacy & security in a connected world. *Washington, DC: Federal Trade Commission.*

35. World Health Organization. (2011). *World report on disability.* World Health Organization.

36. Frauenberger, C. (2015). Disability and technology: A critical realist perspective. In *Proceedings of the 17th International ACM SIGACCESS Conference on Computers & Accessibility* (pp. 89–96).

37. Sik-Lányi, C. (2015). Barriers and Facilitators to Uptake of Assistive Technologies: Summary of a Literature Exploration. *Assistive Technology: Building Bridges, 217*, 350.

38. Pape, T. L. B., Kim, J., & Weiner, B. (2002). The shaping of individual meanings assigned to assistive technology: a review of personal factors. *Disability and rehabilitation, 24*(1–3), 5–20.

39. van den Heuvel, E., Jowitt, F., & McIntyre, A. (2012). Awareness, requirements and barriers to use of Assistive Technology designed to enable independence of people suffering from Dementia (ATD). *Technology and Disability, 24*(2), 139–148.

40. Dijksterhuis, A., & Nordgren, L. F. (2006). A theory of unconscious thought. *Perspectives on Psychological science*, *1*(2), 95–109.
41. Mansell, J. (2006). Deinstitutionalisation and community living: progress, problems and priorities. *Journal of Intellectual and Developmental Disability*, *31*(2), 65–76.
42. Mansell, J., Knapp, M., Beadle-Brown, J., & Beecham, J. (2007). *Deinstitutionalisation and community living–outcomes and costs: report of a European Study. Volume 2: Main Report.* University of Kent. (*106, 112*).
43. Sykes, R. (1975). Housing and Community Development for the Handicapped. *HUD Challenge (March) 1975.*
44. United Nations, Division of Social Policy and Development Disability. Monitoring and Evaluation of Disability-Inclusive Development. (2016). Available online, https://www.un.org/development/desa/disabilities/resources/monitoring-and-evaluation-of-inclusive-development-data-andstatistics.html [Accessed: 10-Oct-2016].
45. United Nations General Assembly (2006). Convention on the Rights of Persons with Disabilities. Geneva, *GA Res, 61,* 106. Available online, https://www.un.org/development/desa/disabilities/convention-on-the-rights-of-persons-with-disabilities.html [Accessed: 11 April 2017].
46. Meeks, L. M., & Jain, N. R. (2015). *The Guide to Assisting Students With Disabilities: Equal Access in Health Science and Professional Education.* Springer Publishing Company.
47. Brown, E. (2016). Disability awareness: The fight for accessibility. *Nature, 532*(7597), 137–139.
48. Björk, E. (2012). *Universal Design Or Modular-Based Design Solutions-A Society Concern.* INTECH Open Access Publisher.
49. World Health Organization. (2001). International Classification of Functioning, *Disability and Health: ICF.* World Health Organization.
50. Mundial, B. (2004). World development report 2004: making services work for poor people. *Banque mondiale,* Washington DC, USA.
51. Maor, D., Currie, J., & Drewry, R. (2011). The effectiveness of assistive technologies for children with special needs: a review of research-based studies. *European Journal of Special Needs Education, 26*(3), 283–298.
52. Jaeger, P. T. (2006). Telecommunications policy and individuals with disabilities: Issues of accessibility and social inclusion in the policy and research agenda. *Telecommunications Policy, 30*(2), 112–124.
53. Timmermans, N. (Ed.). (2005). *The status of sign languages in Europe.* Council of Europe.
54. Kennard, W. E., & Lyle, E. E. (2001). With freedom comes responsibility: ensuring that the next generation of technologies is accessible, usable and affordable. *CommLaw Conspectus, 10,* 5.
55. Goggin, G., & Newell, C. (2003). *Digital disability: The social construction of disability in new media.* Rowman & Littlefield.
56. Cullen, K., Kubitschker, L., Blanck, P., Myhill, W. N., Quinn, G., Donoghue, P. O., & Halverson, R. (2008). Accessibility to ICT products and services by disabled and elderly people. *Towards a Framework for Further Development of EU Legislation or Other Co-ordination Measure on eAccessibility.*
57. Disability Rights Commission. (2004). *The Web: Access and Inclusion for Disabled People; a Formal Investigation.* The Stationery Office.
58. Lawson, A. (2017). The European Union and the Convention on the Rights of Persons with Disabilities: Complexities, Challenges and Opportunities. In *The United Nations Convention on the Rights of Persons with Disabilities* (pp. 61–76). Springer International Publishing.
59. Weber, R. H. (2016). Governance of the Internet of Things—From Infancy to First Attempts of Implementation? *Laws, 5*(3), 28.
60. Weber, R. H., & Studer, E. (2016). Cybersecurity in the Internet of Things: Legal aspects. *Computer Law & Security Review, 32*(5), 715–728.

Chapter 7
Concluding Remarks

The magnitude and experiences that go with the global aging population and a large number of individuals living with dementia or sensory impairments have been detailed in Chap. 2. Enhancing the human experiences for these individuals and their caregivers can be considered as the drivers of the AT, including the IoT-based assistive devices and smart environments. In particular, various studies and the discussions in Sect. 2.2 have clearly shown the links between the views on disability and age on the design, development, and evaluation of AT. Chapters 3–5 have examined the sensing, actuation, intelligence, and the digital revolutions that are behind advanced AT, the IoT, smart environments, and their deployment scenarios for assisting the elderly and those living with dementia or disability. The challenges in moving forward have been considered in Sect. 5.2. Some of the criteria relevant to the usefulness of AT, as well as its performance requirements in the contexts of various regulatory, social, and interactivity frameworks, have been explored in Chap. 6. This chapter concludes this work, summarizing the roles of assistive IoT and smart environments and arguing the case for the need of taking a combinatory and holistic approach for the developments of AT, including assistive IoT and smart environments.

7.1 Assistive IoT and Smart Environments: Where They Stand

There have been many innovative developments and exciting works in the areas relevant to the aged care and improving the quality of life for people with disability or dementia, their families, and their caregivers. This work is focused on the IoT-based AT and smart assistive environments. The last few years have witnessed the deployment of many ICT and some IoT-based related solutions. Chapter 3 presents a small sample of various available or emergent technologies for use by seniors or people living with dementia or disability. Section 5.1 outlines some of the assistive

© Springer International Publishing AG 2017
S. Shahrestani, *Internet of Things and Smart Environments*,
DOI 10.1007/978-3-319-60164-9_7

IoT-based deployments and scenarios. The rapid growth of suitable AT is exciting. However, widespread and effective use of such technologies still seems to be far off. There are still many gaps, requiring major research works and development of appropriate solutions.

A general aspect that needs particular attention is taking a combinatory and holistic approach that can help the enhanced independence, improved access provisions, and integration of the elderly, or individuals living with dementia or disability with all people. This approach can also be of value for the families and caregivers of these individuals. Another major point that needs to considered relevant to deployment of advanced AT is the lack of awareness of how affected people and their families can take advantage of these technologies, sometimes with just some modest alterations. It appears that even many professionals are not aware of available advanced assistive IoT and smart environments. This lack of awareness is one of the key hurdles in effective adoption of these technologies.

Raising awareness can be quite challenging. One of the main difficulties is that in various quarters different age and disability models have led to divergent views on the purposes for AT and the extent of the suitability of available and emergent relevant technologies. The challenges, at least partially, relate to how the medical, healthcare, and other professionals view and identify with the disruptive technologies, such as the IoT, and their usefulness in AT developments. As discussed in Sect. 2.2, the two dominant models of disability have considerable influences on identifying and implementing the various ways that can enrich the lives of the elderly or people with disability or dementia. Such views can also affect the use of technology and the IoT. The salient points of such views, emphasizing what is significant for the deliberations of this work have been summarized in Table 7.1.

The medical model may be seen to be based on pragmatism. It has an apparent purpose for the technology. It aims to utilize technology to identify and then diminish the limitations of the individual with some sensory or cognitive impairment. However, the medical model is based on having a 'norm' for any biological function and considering any deviations from that norm, as a limitation that needs to be corrected. Such normative views can, however, lead to stigma and negative attitudes. The social model is widely accepted for contemporary design and evaluation of AT in the context of diversity and equality. In the social model interpretations, disability is the consequence of a combination of individual-related factors, including the nature of the impairment, stigma and social biases, own and perceived attitudes, economic or environmental barriers, as well as the physical structural problems. Aging can have some or most of these characteristics, in particular with respect to adoption of assistive devices. With these perspectives, the aim of AT is to expand opportunities for people and break down the functional and social barriers, so that the individuals can achieve their goals and aspirations. In this view, different abilities and age are just other categories of difference amongst people. It rejects normative conventions, requiring approaches towards inclusiveness that are similar to those for other categories of difference, like race, sexuality, or gender.

With such a diverse mix of causes, effective solutions and technology developments need to be based on a holistic vision of disability and age. Such a comprehensive view

Table 7.1 Comparison of disability models

	Medical model	Social model
Disability basis	Medical and professional diagnosis for assessments of impairments and deviations of individual's ability from some established 'norm'	Mismatch of opportunities for individuals, who happen to be with some impairment, on equal levels with others
Sees root causes of disability in	Disease, illness, genetics, trauma, or injury	Physical, communication, social, environmental, organizational, and attitudinal barriers that lead to discrimination
Seeks solutions through	Assessment, monitoring, and diagnosis of the impairment through medical diagnosis, then • Curing the individual, considering the economic factors, or • Tolerating the 'abnormality' and providing the necessary care to support the incurable individuals • Intervention provided and controlled by professionals and experts	Removal of physical, communication, social, environmental, organizational, and attitudinal barriers
Advantages	Curing the physical and mental impairment or relieving the associated pain	Eliminating discrimination and promoting inclusion, through identifying the adjustments, needed to be made to devices, services, communications, buildings, transport systems, educational institutions, and workplaces, among others
Issues and shortcomings	• Paternalistic and protective approach • Focus on cure or care, rather than on the person • May justify segregation and institutionalization • May fosters prejudices by relating discrimination to medical conditions • Rejected by many individuals with disability and related advocacy groups • Ignores the full range of disability experience • May overlook the abilities of people with some impairment • May put physicians and other healthcare professionals in charge of decision-making for capable individuals	• May burden the society, particularly with the aging population that may give rise to the numbers of people with some impairment • Maybe at odds with the views of dedicated healthcare professionals • May amount to denying or dismissing the personal experience of physical and intellectual limitations of the underlying disease or injury

has to embed many aspects. For instance, as discussed in Chap. 3, AT developments can benefit from many advances in related diverse areas, including psychology, industrial design, AI, robotics, computer vision, sensors, wireless communications, medicine, traditional AT, urban planning, autonomous vehicles, and many others. These combinations can help with increasing the adoption of AT by their anticipated users.

A holistic vision that is based on the social model can credibly identify various ways that enrich the lives of the elderly and people with disability. The vision can be used, for instance, to address the integration of an individual in educational settings. The inclusion may be contingent on regulations, like anti-discrimination legislation at the social level, and on specialized devices or services at the individual level. This view can also lead to the development of technologies and devices that are suitable for a range of people regardless of their age, gender, or abilities. Such developments meet the design for all concepts that as highlighted in Sects. 2.3 and 6.3, can significantly affect AT adoption and employment in positive manners.

Another aspect of the holistic vision for disability and aged care relates to the necessity of engaging various stakeholders. The stakeholders include healthcare professionals, affected individuals and their families or caregivers, technology developers, policy makers, and even the general community in some cases. Without such engagements, the solutions and interactions with people may ignore some features that may be considered to be insignificant by even a committed stakeholder, making the solution or technology ineffective or inefficient, as far as others are concerned. The best solutions can only be beneficial to the extent that they are adopted and used. As such, functionality and user acceptance are both of vital importance. The acceptance of a device or technology is a complex concept that depends on several characteristics, including functionality, how well they address the impairment-related issues, ease of use, user-friendly interfaces, as well as aesthetics, social and cultural approval, and facilitation of inclusion. As mentioned before, eyeglasses and contact lenses provide good examples of the AT that meet these criteria. They are accepted so much that most people do not see them as assistive devices or technology. Some people even use them as statements of fashion for expressing themselves.

Crafting personalized experiences for individuals is also another important consideration for AT adoption and employment, as outlined in Sects. 3.2 and 5.1. Such an aspect can be very beneficial in aged care and particularly for people with dementia or some other forms of cognitive impairment. As mentioned at the beginning of Sect. 6.1, some people may feel comfortable and function effectively in their personal and familiar spaces, as they may have developed a sense of running on autopilot due to their personal familiarities. Smartphones, several other usual devices, and definitely the IoT and smart environments can provide the essential components for such personalized experiences. As substantiated in Chap. 4 and Sect. 5.1, smartphones, communication technologies, and many other tools and devices that provide the required capabilities for assistive IoT are either already available or emergent. However, developments of practical solutions and suitable AT that their anticipated users adopt and widely employ require substantial work and perhaps more importantly, public and private investments. With the aging population and the massive number of people living with some impairment, detailed in Sect. 2.1, as well as the scale and growth rates of the underlying technologies, assistive IoT and smart environments deserve to receive considerably more targeted and yet holistic and combinatory research investigations. The massive market, progressive regulatory frameworks, considered in Sect. 6.4, technological and societal advances, and most importantly, taking a holistic view can provide transformative solutions to enhance the quality of life for the elderly, and the people living with some sensory or cognitive impairment. Developments and adoption of such solutions also depend on having clear visions and being able to engage various stakeholders and the society as a whole, in a continuing quest for new technologies relevant to the assistive IoT and smart environments.

7.2   Assistive IoT and Smart Environments: Where to from Here

The IoT has provided real solutions to many problems. It has the potential to significantly improve aged care, healthcare, and quality of life for people with sensory or cognitive impairment. The IoT sensors may even be superior to human senses of hearing or sight. The IoT can sense its environment continually at low costs and communicate with other machines or provide information to humans through smartphones and similar devices. The key points relevant to making these progresses, are network architectures and platforms, sensor technologies and miniaturization, development of services and applications, usability and interoperability, and addressing the privacy and security concerns.

For the right or the wrong reasons, all new phenomena and technologies, including assistive IoT and smart environments are subject to controversial views. While the IoT, wearable devices, and smart environments have resulted in the development of many new or enhanced devices and services for an assisted living, the adoptions of these products have not always been high. One of the issues, in many cases, is the lack of a holistic vision to guide the development and integration of such products and services with the existing and widely used AT. As it stands, with the very fast pace of expansions and growth, many of the devices or services seem to be developed in isolation, with connectivity and integration being after-thoughts.

Some people may feel intimidated by the newer technologies such as those using smart environments and the assistive IoT. A similar issue arises from the fast pace of developments in such advanced technologies. The costs, learning curves, or simply the lack of awareness can also mean that many seniors or people with disability may simply not use the new technologies. Some may not even try to identify newer AT, such as those based on the IoT, for the older people who may be viewed as being reluctant to engage with the modern devices and services.

Smartphones and the IoT can provide the enterprises with the essential material for crafting personalized experiences for their clients to achieve better outcomes. This individualization can be particularly an important feature for aged care and supporting people with dementia. However, to benefit from that feature, more advanced and less-intrusive interfaces, such as wearable smart devices or smart environments, and an understanding of their interactions with other entities in the ecosystem, are crucial. Other advanced technologies such as interactive IoT-based smart environments, various cloud-based services, a diverse range of communication technologies, big data analytics, and development of the appropriate related apps for smartphone, also have vital roles to play.

The smart and assistive technologies that have been outlined in this work can have far-reaching life changing values in expanding access for the elderly and people with some sensory or cognitive impairment. To take these transformative IoT-based solutions into reality, however, either the private sector must be able to make monetary profits from them or activism, social prosperity, and governments need to lead the way.

Based on the contents and deliberations of this book and the literature cited in it, with regard to future directions for work in the areas of assistive IoT and smart environments, the following points can be made.

- The unique insights of the elderly and people living with disability or dementia, as well as their informal and formal caregivers, make it essential to include and involve them in various stages of AT developments.
- It should be noted that the evaluation of a device or service may produce dissimilar outcomes for different stakeholders. For example, policymakers may find cost-savings associated with some technology a positive aspect, while the users or the caregivers may not be much concerned about that. Therefore, all stakeholders need to be included and involved in all steps of AT and assistive IoT developments. Stakeholders may include healthcare professionals, researchers and academics, professional developers, designers, manufacturers, service providers, relevant advocacy groups and organizations, government agencies, and policy makers.
- The research and developments must be multidisciplinary, combinatory, and based on holistic visions. A holistic approach needs to consider technical aspects, experiences of various stakeholders, social acceptance, avoidance of stigma, and other human elements.
- To develop assistive IoT-based systems and AT that are actually adopted, a holistic approach that incorporates the mental state and wishes of the end users is required. Otherwise, the resulting products may only serve a perceived or assumed need, rather than filling an actual gap.
- The combination of big data analytics and the IoT provides assistive living solutions that can be holistic in nature. However, the overwhelming amount of data can also disenfranchise many individuals. Many people may find it hard to see how to make the best use of the available information. This difficulties may even include the selection of appropriate devices and services that should meet their needs.

- The use of advanced technologies must be justified based on the assistance that they actually provide, rather than on just being high tech or new. Expenses, ease of use, availability, and the languages used are among the factors that play significant roles in adopting an AT solution by its intended users.
- To succeed in providing access and inclusion, aesthetics and affordability of AT and assistive IoT are as important as their functionalities.
- More investigations into what interfaces produce the expected performance and how the human-device interactions should be implemented are needed. Published works on how the end users prefer to interact with the AT in general, and more specifically with assistive IoT, seem to be rather sparse and sketchy.
- A valid or at least widely accepted approach that can correlate between the results of lab-based experiments and real-world experiences of the elderly or people living with dementia or some impairment is required. The need for such validations is particularly the case, for assistive IoT and smart environments that may be pervasive in nature and use many objects with diverse characteristics and functions.
- A comprehensive investigation and systematic review of the results of the previous studies on the factors that influence the adoption and employment of AT, and specifically assistive IoT and smart environments, is needed.
- Sensors and actuators must function unobtrusively. Wearables, additionally, need to be comfortable to use.
- Context awareness and environmental surveillance of smart environments need to be adaptive, capable of addressing the evolving changes, as well as the unexpected events and emergencies.
- Assistive IoT-based systems and smart environments should be able to adapt to emerging tools and mechanisms, such as new sensors, other connected objects, deep learning, and communication technologies, without too much cost or difficulty.
- Lack of awareness of available IoT-based AT is a major obstacle to their adoption. Many newer AT and related technological developments share the same faith. End users and many professionals may not be aware of what is already available. The lack of awareness can be even more pronounced for the cases that with minor modifications of existing products, they can conveniently address many identified needs and issues. Raising awareness, in different corners, is a necessity.
- The regulatory frameworks and the policies relevant to the developments of AT must aim to lessen the social misperceptions about age and disability and alleviate the effects of misconceptions on individuals. Development of devices and services that work with mainstream technologies or make them more accessible can be of value for this aim. Such developments should be part of a holistic approach to assistive IoT and smart environments.
- In some cases, to allow the rapid and innovative growth of the IoT, and also more generally for the ICT advancements, the regulatory bodies may overlook the accessibility requirements. The diversity and prevalence of ICT and the services that are provided by them may make it impossible or impractical to legislate for

all potential cases. To address the IoT and ICT accessibility, smarter regulatory frameworks need to be developed and enforced.

- Many people are wary of the invasion of their privacy by the IoT or connected devices and appliances. Even perceived risks to privacy and security, whether realized or not, can easily damage the confidence of the end users, leading to lowering of the adoption and employment of these technologies. Security and privacy issues, particularly relevant to personal information, need to be addressed. However, it can be noted that dealing with these matters for assistive IoT-based systems and smart environments can be challenging.

- AT and assistive IoT should be developed based on universal design or design for all concepts. To reduce stigma, the technology in use should be the same for all, when possible. Misperceptions about their abilities can be mitigated if people with different abilities employ the technology that everyone is using. Universal design can help with social acceptance and mitigating stigmatization by minimizing the differences between AT and their mainstream counterparts, improving the chances of their adoption. The IoT and smart environments can be of significant value in this respect.

Computers, smartphones and other smart devices, and to some extent some other electronic devices can be particularly suitable for use as part of or in conjunction with other AT. They are also widely used by many people. The IoT, smart homes, buildings, cities, and other smart environments appear to be also expanding rapidly. Familiar devices, like smartphones and their included gadgets, and smartwatches or smart TVs and other smart appliances can provide the required interfaces for interactions with the intended end users of the assistive IoT. The associated technologies can be of all-inclusive forms, acceptable to most groups of people. As such, with their potentially ubiquitous nature and large growths, the IoT and smart environments can and should play leading roles in the aged care and improving the lives of people with disability or dementia.

Index

© Springer International Publishing AG 2017
S. Shahrestani, *Internet of Things and Smart Environments*,
DOI 10.1007/978-3-319-60164-9

Printed in the United States
By Bookmasters